COMMUNITIES IN TRANSITION: PROTECTED NATURE AND LOCAL PEOPLE IN EASTERN AND CENTRAL EUROPE

Ashgate Studies in Environmental Policy and Practice

Series Editor: Adrian McDonald, University of Leeds, UK

Based on the Avebury Studies in Green Research series, this wide-ranging series still covers all aspects of research into environmental change and development. It will now focus primarily on environmental policy, management and implications (such as effects on agriculture, lifestyle, health etc.), and includes both innovative theoretical research and international practical case studies.

Also in the series

Sustainability and Short-term Policies
Improving Governance in Spatial Policy Interventions
Edited by Stefan Sjöblom, Kjell Andersson, Terry Marsden and Sarah Skerratt
ISBN 978 1 4094 4677 4

Energy Access, Poverty, and Development
The Governance of Small-Scale Renewable Energy in Developing Asia
Benjamin K. Sovacool and Ira Martina Drupady
ISBN 978 1 4094 4113 7

Tropical Wetland Management
The South-American Pantanal and the International Experience
Edited by Antonio Augusto Rossotto Ioris
ISBN 978 1 4094 1878 8

Rethinking Climate Change Research
Clean Technology, Culture and Communication
Edited by Pernille Almlund, Per Homann Jespersen and Søren Riis
ISBN 978 1 4094 2866 4

A New Agenda for Sustainability
Edited by Kurt Aagaard Nielsen, Bo Elling, Maria Figueroa and Erling Jelsøe
ISBN 978 0 7546 7976 9

At the Margins of Planning
Offshore Wind Farms in the United Kingdom
Stephen A. Jay
ISBN 978 0 7546 7196 1

Communities in Transition: Protected Nature and Local People in Eastern and Central Europe

SASKA PETROVA
University of Manchester, UK

ASHGATE

Published by
Ashgate Publishing Limited
Wey Court East
Union Road
Farnham
Surrey, GU9 7PT
England

Ashgate Publishing Company
110 Cherry Street
Suite 3-1
Burlington, VT 05401-3818
USA

www.ashgate.com

British Library Cataloguing in Publication Data
A catalogue record for this book is available from the British Library

The Library of Congress has cataloged the printed edition as follows:
Petrova, Saska.
  Communities in transition : protected nature and local people in Eastern and Central Europe / by Saska Petrova.
      pages cm. -- (Ashgate studies in environmental policy and practice)
  Includes bibliographical references and index.
   ISBN 978-1-4094-4850-1 (hardback) -- ISBN 978-1-4094-4851-8 (ebook) -- ISBN 978-1-4724-0183-0 (epub) 1. National parks and reserves--Europe, Central--Management. 2. National parks and reserves--Europe, Eastern--Management. 3. National parks and reserves--Europe, Central--Citizen participation. 4. National parks and reserves--Europe, Eastern--Citizen participation. 5. National parks and reserves--Social aspects--Europe, Central. 6. National parks and reserves--Social aspects--Europe, Eastern. 7. National parks and reserves--Political aspects--Europe, Central. 8. National parks and reserves--Political aspects--Europe, Eastern. 9. Natural resources--Europe, Central--Management. 10. Natural resources--Europe, Eastern--Management.. I. Title. II. Title: Protected nature and local people in Eastern and Central Europe.
  SB484.E8P48 2014
  363.6'8--dc23
                                                                    2014005031

ISBN 9781409448501 (hbk)
ISBN 9781409448518 (ebk – PDF)
ISBN 9781472401830 (ebk – ePUB)

Printed in the United Kingdom by Henry Ling Limited, at the Dorset Press, Dorchester, DT1 1HD

Contents

List of Figures

List of Tables

To Stefan

Preface

North America's vast national parks, Africa's complex wilderness reserves and Asia's precarious rainforest biomes are some of the main associations invoked by the notion of 'nature protection' in most developed-world debates on the subject. The suggestion that Europe can be of major relevance in this regard has traditionally been outside the radar of the public and academic mainstream, especially when referring to the post-communist countries of Central and Eastern Europe (CEE). Yet the 22 states that occupy the territory between the Baltics and the Balkans possess a sizeable network of protected areas, underpinned by high levels of biodiversity, substantial forest cover, low population densities and intensifying processes of rural abandonment.

The nature conservation opportunities afforded by this emergent socio-ecological setting have not gone unnoticed by nature conservationists and the media. In 2009, for example, the BBC produced a documentary titled 'Iron Curtain, Ribbon of Life' where a region that otherwise bustled with life during socialism was portrayed as 'a forgotten backwater'. Such thinking demonstrated that David Attenborough-style filming of 'natural' and 'empty' landscapes can now be undertaken in the heart of Europe, not far from some of the very studios where the programmes are edited and produced. It reflected a growing romanticization of areas neighbouring the Iron Curtain as a new 'European Green Belt' in which, just like many Attenborough documentaries, humans are airbrushed from the picture in discussions about the functioning and protection of ecosystems.

This monograph, therefore, is motivated by the need to address the relative neglect and misapprehension of the role of local people in contemporary debates on nature conservation. I also wish to highlight some of the more generic issues situated at the nature-society nexus within national parks in the transitional context of CEE, where dynamics of socio-political restructuring have been followed by fundamental changes in environmental governance practices. The post-1990 reconfiguration of conservation paradigms in this part of the world has re-opened various age-old debates about the protection and administration of natural phenomena selected for such purposes. Further complicating the situation has been the introduction of market-based principles, which has embedded the entire process in broader dynamics of neoliberalization and the capitalist space economy. Although state officials have recognized the necessity to involve local people in their work, the manner in which such groups are being defined and included in decision-making agendas remains uncertain and highly contested. The human inhabitants of protected areas are often presented as a monolithic entity with a singular set of interests, rather than a complex assemblage of individuals whose desires and aspirations are nuanced, diverse and intertwined with the non-human world.

The book aims to provide an integrated perspective on why, how and for whom nature conservation practices have been implemented in CEE, in addition to shedding further light upon the mechanisms through which such practices both redefine and are affected by the everyday life of people living in national parks. This is underpinned by a desire to think about the establishment and functioning of national parks in political ecology terms. The book thus provides a critical global review of the environmental motivations and power interests behind the creation of national parks, as well as a typology of the relations between local people and dynamics of nature protection in them. In situating my analysis within a global framework, I wish to challenge the dichotomy between developed and developing countries that pervades much of the academic literature on nature protection. I also explore the multiple meanings and practices of 'community' within protected areas, so as to underline the political subjectivities that come to the fore when conceptualizing such formations as fundamentally fluid and relational.

I would have been unable to undertake the background work for this book without the financial support of the Czech Ministry for Education – for which I am particularly grateful – as well as the three higher education institutions where I have been based during the period required to prepare and write it: the Faculty of Natural Sciences at Charles University in Prague (2007–2011), the School of Geography, Earth and Environmental Sciences at the University of Birmingham (2011–2012), as well as the School of Environment, Education and Development at the University of Manchester (since 2013). Throughout this time, I have benefited from discussions and interactions with colleagues working in cognate areas, including Martin Braniš[†], Martin Čihař and Ludek Sýkora (Prague), Stefanka Hadzi Pecova (Skopje), Iwona Sagan (Gdańsk), Michael Gilek, Michael Gentile and Kathryn Furlong (Stockholm), Rosie Day, Nikoleta Jones, Oleg Golubchikov and Dominique Moran (Birmingham), Vanesa Castan Broto (London), Mike Bradshaw and Jon Coaffee (Warwick).

This book would not have been completed without the extremely supportive environment provided by the geography discipline at Manchester led by Martin Evans, and colleagues such as Tim Allott, Noel Castree, James Evans, Simon Guy and Kevin Ward. I am also grateful to my dear friends and colleagues Maria Kaika and Erik Swyngedouw for the benefits accrued from our vigorous and inspirational discussions.

The background work towards the book would have been impossible without the assistance of the management authorities of, and rangers from, the Pelister and Šumava National Parks, the local people included in the research and the Civic Association Biosfera with Nesad Azamovski from Bitola, Macedonia. My parents (Dafinka and Petre), sister (Ivana) and parents-in-law (Eleni and Dimitrije) provided unreserved love and support throughout the vicissitudes that necessarily accompany an undertaking of this size. My foremost gratitude is reserved for them.

This book is dedicated to my husband Stefan, without whom all of this would have been impossible.

Saska Petrova
Shtip/Skopje/Manchester
January 2014

Chapter 1

Introduction

Nature protection has never been a straightforward and simple process; ever since the establishment of the world's first protected areas more than a century ago, state authorities responsible for the management of such spaces have been constantly forced to reconcile overlapping interests and priorities in order to achieve their goals. In recent decades, the governance of protected areas has been further complicated by the rapid urgency of the need to conserve the Earth's rapidly diminishing biodiversity stock, as well as the challenges associated with climate change mitigation and adaptation (Munck 2005, Peck et al. 2010, Bakker 2010, Castree 2011). Such developments have been accompanied by wider efforts to expand, formalize and intensify the rate and extent of market-based transactions, commonly encapsulated under the term 'neoliberalism'.

Natural resources located within or near protected areas have become targets for the expansion of the economic and political agendas associated with this process. The extensive social implications of neoliberalism across the world have highlighted the need for establishing new poverty alleviation mechanisms, thus placing further pressure on those seeking to balance social and environmental demands in the sustainable development of protected areas. However, the micro-scale effectiveness of management practices in protected areas has often hinged on the manner in which the residents of such spaces relate to the governance practices, institutional structures and political discourses that frame nature conservation. It has become clear that protected area governance often plays a significant role in the rise of 'actually existing neoliberalism' – commonly termed 'neoliberalization' – with respect to the articulation of everyday life.

At this point, it should be emphasized that the last five decades have seen the publication of a significant body of academic and policy-orientated scholarship aimed at unravelling the multiple political and social aspects of local community participation in protected area governance (Kapoor 2001). But much less has been written about the manner in which different nature protection regimes have shaped the perceptions of protected areas as management institutions, on the one hand, and inhabited places with a specific material and imagined geography, on the other. In particular, it remains unclear to what extent 'the crisscrossing of social relations, of broad historical shifts and the continually altering spatialities of the daily lives of individuals' (Massey 2001: 462) has impacted the construction of attitudes towards nature protection policies in such areas.

As protected areas with specific symbolic, political and historical connotations, national parks both embody and expose the multiple tensions associated with nature protection efforts in developed and developing countries alike. Yet there is a limited amount of knowledge about the community-level relationships between everyday life, neoliberalism and protected area governance in national parks. This is despite the frequent calls, from a range of social scientists, that scholarship on neoliberalism needs to make the connections between top-down projects and everyday life across a variety of scales and spaces (Barnett 2005).

The relative neglect of small-scale socio-political relations underpinning the existence of people in parks is particularly pronounced in the post-communist states of Central and Eastern Europe (CEE), which have undergone major structural changes over the past 20 years. This part of the world has seen the sudden replacement of the centrally planned economy and the one-party political system with a market-based, multi-party democracy. Many countries in the region have joined the European Union (EU), adding a new set of diverse policy practices to an uneven regional landscape affected by the concurrent spread of neoliberalism. The management of protected areas has moved from a hierarchical, tightly-controlled and often inefficient governance framework into a fragmented variety of locally-contingent regulatory situations, where non-state actors play an ever-increasing role.

Research on the relationship between local people and national parks in this region is particularly rare: very little is known about the political, economic and social agencies involved in shaping the everyday lives of the residents of such territories. This is despite the fact the findings of academic research which emphasize that protected areas in CEE are characterized by conflicts between centralized decision making and new public values and concerns (Cellarius 2004, Staddon 2009, Petrova et al. 2009), while lacking the civil society and resource management structures that characterize societies with established nature conservation models (Grujičić et al. 2008, Rodela and Udovč 2008, Švajda 2008, Lawrence 2008).

Purpose of the Book

In light of the knowledge gaps highlighted above, this monograph explores the relationship between local people and nature protection in the 'transition' context of CEE. Using evidence drawn from a range of theoretical and empirical investigations – including on-site research in the region that has involved interviews, questionnaires and surveys of legal and policy documents – the book connects the attitudes of local residents towards protected areas with the history and politics of public participation and environmental governance.

One of the major themes in the book is the extent to which, as discussed by Vorkinn and Riese (2001: 252) in previous work on the subject, 'the values reflected through place attachment' influence the local residents' views of national park management practices. Attention is drawn to the divergent attitudes and perceptions exhibited by local residents in terms of issues such as place affiliation, the state of the environment, nature protection, environmental management, job opportunities and tourism in national parks. By investigating how the residents' links to the places in which they live relate to their appraisal of the national parks' existence – as well as the quality of nature protection and the management institutions of these areas – I seek to determine the kinds of conditions under which local people may develop a negative or positive attitude towards the presence of national parks.

Much of the academic literature that formed the starting point for the research leading to this book sees protected areas as geographical settings for the conduct of everyday life. Therefore, a significant part of the evidence and discussion presented in the text that follows is focused on investigating local people's[1] participation in, and understanding of, a range of national park operation and governance aspects as they relate to the 'geographies of quotidian' (Hubbard and Holloway 2000). Some of the main questions I ask include:

- How is the governance of national parks contingent upon the ability of broader political relations, socio-cultural aspirations and economic practices to shape the course of nature protection approaches?
- In what manner do local people in national parks view the opportunities afforded by such areas, particularly in terms of improving the quality of their everyday lives? How do local attitudes, perceptions and aspirations vary across social and spatial settings when it comes to nature protection?
- Does the degree of place attachment have an influence on local people's perceptions of national parks? How are notions of community affiliation and participation articulated in relation to the process of nature protection?

In a broader sense, the book uses these questions to explore the extent to which nature protection is an obstacle for local economic growth in national parks, as opposed to views that the designation of a national park can be seen as an opportunity for the development of local communities, and a source of financial capital for the protection of the parks themselves. I am also interested in the mechanisms that allow neoliberalism to be 'challenged, resisted and changed by its encounter with nature' (Castree, Duffy and Moore 2010: 744, Bakker 2010) in the context of protected area management. Part of my discussion is aimed at exploring the lessons that can be learnt by applying the experiences of community participation in environmental management in CEE to other locations undergoing major systemic change in their environmental governance practices, such as the 'low carbon transition' that is currently unfolding at a global scale.

The book employs a variety of strands of knowledge in studying such questions. In a number of instances, I draw on the wide body of academic research that has

---

1   In light of the numerous controversies surrounding the delineation of nativity and local residence (Mulder and Coppolillo 2005) I equate the syntagm 'local people' with the standard definition of the term 'local resident' provided by the Oxford English dictionary: 'a person who lives somewhere permanently or on a long-term basis'.

explored the ways in which the geographies of everyday are implicated in the domestication of neoliberalism. Inspired by Stark (1996) and Watts (1998), I also rely on the notion of 'recombinant capitalism' to explore the institutional context of protected area management, while investigating how environmental governance has been implicated in dynamics of privatization, liberalization and corruption. It should also be pointed out that the book, in many ways, builds and develops on the findings of my previous work on nature protection in the CEE context, some of which is based on the same empirical corpus (Petrova et al. 2009a, Petrova et al. 2009b, Petrova et al. 2011). Before providing a more detailed explanation of the theoretical framework of the book and its chapter structure, I would like to briefly outline some of the key debates relevant to my arguments, as well as the case studies explored in the book.

Local People and Nature Protection: Challenges Associated with Local Community Participation

Local populations play an important role in allowing protected areas to be managed in an effective and durable manner. They have mainly been implicated in the governance of national parks via 'communities of place'. However, communities of interest are also of relevance here; especially since the underlying dynamics associated with the articulation of relevant economic, social and political relations extend well beyond the physical spatial boundaries of protected areas. Both understandings of community are further defined and discussed in Chapter 2.

Traditionally, local populations and communities were seen as obstacles towards the sustainable management of national parks, as a result of their allegedly negative attitude towards nature protection. Classic conservationists argued that local people cause a serious threat to the biological integrity of national parks, mainly due to the need for exploiting the natural resources of such areas for livelihood purposes (Terborgh 2004). The so-called 'Yellowstone model' provides one of the most prominent material manifestations of such thinking. It is based on the management approach adopted in creating and governing Yellowstone National Park – the first such area in the world, designated as early as 1872 – which was used as the initial blueprint for protected area planning and management across the world (Stevens 1997). The policies associated with this paradigm entail a biocentric management approach towards nature conservation, including the displacement and exclusion of local communities from designated national park areas. Its underlying tenets see the human disruption of 'natural' environments as a causal factor for the depletion of geological and biological diversity (Ghimire and Pimbert 1997).

'Yellowstone' has been held to account for the high costs of park protection, the lack of success of nature conservation projects, and the neglect of local people's needs and interests across the globe (Brockington and Igoe 2006). Its failures have prompted a range of polarized debates regarding the relationship between local communities and national park management. Most discussions on the topic have revolved around the broader implications of policies that involve curtailments of the economic activities and everyday life of local people (Pimbert and Pretty 1995). One of the most commonly-heard arguments revolves around the existence of a mutually-reinforcing relationship between the socio-economic and political exclusion of the residents of protected areas, on the one hand, and the rise of local opposition leading to the obstruction of conservation objectives, on the other. Basically, authors working in this vein have emphasized that the prioritization of biodiversity conservation has resulted in the relative neglect of the needs, interests and aspirations of local people living in such areas (Duffy 2010). It is now commonly accepted that conventional conservation practices have generated the displacement of local populations from protected areas, while placing unreasonable limitations on the use of natural resources in the name of nature protection (Stevens 1997, Kusumanto and Sirait 2001).

As was pointed out above, the last two decades have led to an increasing realization that local involvement in protected area governance is a key necessity for the sustainable and efficient protection of wildlife, and the articulation of economically effective methods for the everyday care and protection of the environment (Pimbert and Pretty 1995). In order to move beyond declarative commitments, a range of 'co-management' models have been developed in order to encourage the active participation of local people in the governance of nature protection through a flexible and inclusive approach. For example, Community-Based Natural Resource Management (CBNRM) has been comprehensively promoted in recent years as a successful technique for the reconciliation of nature conservation and socio-economic objectives via context-based incentives to local people (Kellert et al. 2000, Büscher and Dressler 2012). As argued by Hall (1999) international development agencies have widely accepted co-management approaches in the governance of natural resources, thus allowing for the extensive involvement of state institutions at the national and local level, as well as civil society organizations and local communities. At the same time, the recognition of local participation has led to the creation of new roles for conservation professionals and protected area authorities, requiring a fundamental re-framing of existing co-management schemes and inter-institutional arrangements (Wells and Brandon 1992, McShane and Wells 2004), in addition to fostering a deeper understanding of ecosystem dynamics.

Despite the profusion of academic research – and numerous examples of local involvement in the running of protected areas – there is still limited evidence that 'community participation has become widespread practice or been effective in influencing the nature and scale of development' (Goodall and Stabler 2000: 63). As pointed out by Pimbert and Pretty (1995: 34) 'the professional challenge for protected area management is to replace the top-down, standardised, simplified, rigid and short-term with local-level diversified, complicating, flexible, unregulated and long-term natural resource management practices'. This is because, in its entirety, community participation opens the opportunity for the incorporation of 'widely different levels and qualities of involvement at the local level' (Pretty 1995: 4). Such arguments have been further supported by Defries et al. (2007), who emphasize that the extent and magnitude of human resource use in a protected area management system is related to its achievement of an effective balance between human needs and ecological functions. According to Daily and Ellison (2002) and Rosenzweig (2003), the main purpose of protected area management should be the establishment of 'win-win' solutions that satisfy human needs while maintaining ecological functions.

Adding to the patchy evidence of democratic local participation in nature protection – a theme that is explored in this book through both literature reviews and primary data – is a second discrepancy between theory and practice. Most of the existing models for local community involvement in protected area management have been based on the experiences of relatively stable and developed societies (Hall 2000). Developing world countries still host controversial debates about the extent to which nature conservation practices should follow the 'Yellowstone' model – excluding any 'interference' from local people – as opposed to approaches that would accommodate development and livelihood needs at the same time (Stevens 1997). This is especially true in spaces where protected area management may conflict with community-based conservation and 'nature-aware' tourism. In practice, many developing-world countries still overwhelmingly utilise top-down management models, partly as a result of the chronic lack of research and policy awareness relating to the ecological, socio-political and economic factors that influence effective community engagement in protected area governance, biodiversity conservation, and rural economic development more generally (McShane and Wells 2004).

Case Study Research: People and Parks in Transition

The book's explorations of the multifaceted views of local residents and communities with respect to the governance and features of national parks are undertaken at two levels: i) globally, with the aid of a review of existing published research on the subject; and ii) locally and nationally in CEE, where I focus on a detailed exploration of nature protection approaches as they are formulated and practiced in Macedonia and the Czech Republic as representatives of, respectively, a developing country in south-eastern Europe and a developed, EU-member Central European country. In addition to surveys of national-scale legal and policy documents, these analyses are mostly based on evidence drawn from in-depth expert and household interviews as well as standardized questionnaire surveys undertaken in Pelister National Park (Macedonia) and Šumava National Park (Czech Republic) between 2008 and 2009 (see figures 1.1, 1.2 and 1.3).

The choice of case study countries and national parks stems from the combination of similarities and differences that they contain with respect to physical and socio-demographic characteristics, as well as the influence of wider

Figure 1.1   The case study countries in their broader geographical context
Note: The locations of figures 1.2 and 1.3 are indicated by the letters 'A' and 'B' respectively.

Figure 1.2   Salient features of Pelister National Park

Note: Tourism case-study sampling point is represented by two concentric circles. Black lines represent national park and/or state boundaries.

processes such as the post-communist transition and EU integration. The case studies are located in a continent that has been subject to long-standing processes of human settlement and industrialization. These dynamics, which have mostly been concentrated in Europe's central and northern parts, have combined with

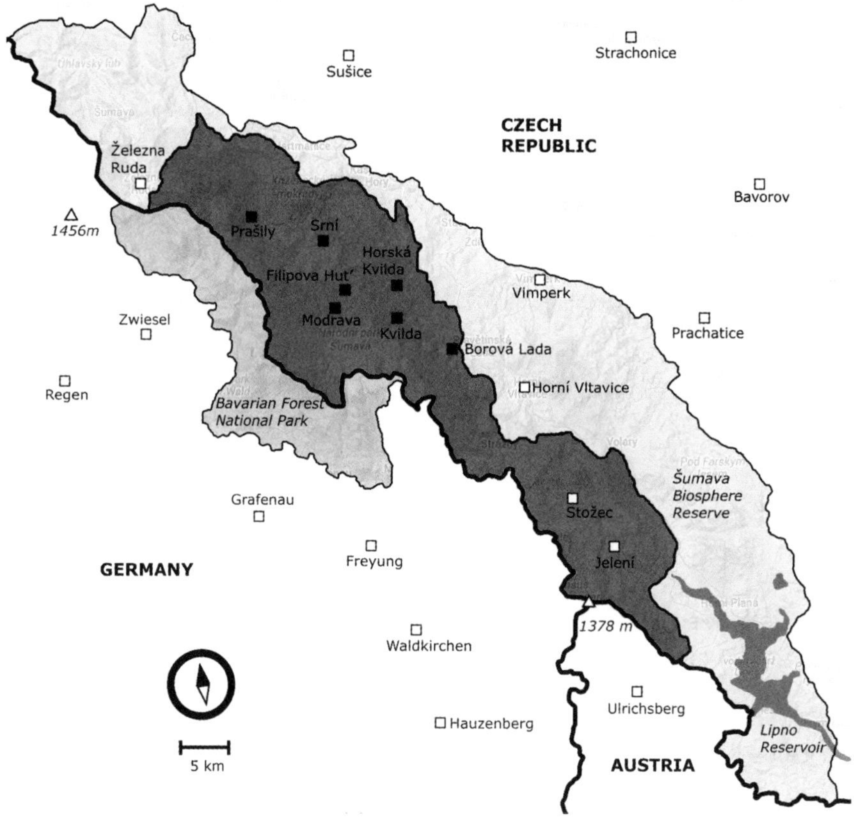

Figure 1.3   Salient features of Šumava National Park

Note: National park territory indicated by the darkest shade of grey.

antecedent physical geography patterns to create a situation wherein most of the continent's unique flora, fauna and biogeographic zones are located in its eastern, southern and central regions. Such spaces still host a number of pristine habitats: from forest remnants in the Carpathians (which may serve as baseline research areas for ecology, see Wesołowski 2007), to formerly-restricted areas along the former Iron Curtain that harbour high levels of biodiversity thanks to experiencing little human influence for more than five decades.

CEE, however, has recently undergone a rapid process of economic restructuring, resulting in far-reaching environmental transformations. The agricultural and forestry sectors provide numerous examples of such systemic changes (Báldi and Faragó 2007). During communist central planning, farming in eastern Europe was just as intensive – at least in terms of fertilizer use – as in other developed European countries. The post-1990 collapse of large, state-managed agricultural enterprises resulted in a decrease in cultivation intensity and fertilizer use, which had positive effects on farmland biodiversity. Nevertheless, grasslands suffered because of the abandonment of grazing, and forests suffered because of the intensification of logging. Even large, protected forests such as the famous Bialowieza Forest were not immune from felling concessions (Wesołowski 2005).

The management of protected areas has also reflected the socio-economic and political changes experienced by the region. Thanks to their EU membership aspirations, both Macedonia and the Czech Republic instituted radical changes to their environmental legislation, involving an entire new array of nature protection laws and regulation. Yet the disparate speed and scope of the accession process – the Czech Republic joined the EU in 2004, while Macedonia is still a candidate country – have combined with wider economic, political and spatial processes to create fundamental differences in the two countries' nature protection regimes. The two national parks covered by this book, therefore, operate within relatively dissimilar state policy settings. In should also be noted that Pelister and Šumava diverge in terms of their local socio-economic history, land-use practices and natural features.

What makes the two areas similar, however, is the fact that both of them can be found in mountainous, forest regions, situated on the border with Greece in the case of the former, and Germany and Austria in the case of the latter. As such, the parks lie adjacent to similar protected spaces in neighbouring countries. Both of them have been designated at the state level and are managed by hierarchically organized and rigidly-structured authorities, strictly controlled by the respective central governments. Pelister and Šumava alike are second category-protected areas according to International Union for the Conservation of Nature's classification, and as such are meant to contain – formally at least – similar zoning patterns and management practices.

Given such contingencies, my explorations of the relationship between nature protection and local communities have aimed to encompass – to the greatest extent possible – the wider social, economic and environmental circumstances of the two parks. Thus, the research questions listed in the previous section were investigated comparatively for both parks and with the aid of a mixed approach (Winchester 1999, Kwan 2004) that included both qualitative and quantitative analyses. One of the mainstays of my research was a standardized questionnaire survey, involving 131 and 182 households in Pelister and Šumava, respectively. It was executed during the summer of 2009 in the case of the former, and 2008 in the case of the latter. I targeted the questionnaires towards a random sample of local villagers and second-home owners older than 16 years of age in six (in the case of Šumava) and three (in Pelister) rural settlements situated in or around the parks (see figures 1.2 and 1.3). As a result of the specific nature of Pelister National Park, one of the surveyed villages was located 1 km outside its boundaries, although forest and agricultural land belonging to its inhabitants lies within the park.

Since the selected villages are relatively small, every single house was visited as part of the sampling procedure; the visit would not be repeated if there was no response at the door. Only one person per household was interviewed – usually the one who answered the door. The interviews lasted between 15 and 20 minutes, and took place inside the respondents' homes (since many authors consider interviewing respondents in their domestic environment as a relative advantage in this regard (see Disman 1993, Nováková 2004, Mundilová 2007). The response rate in Pelister was 97 per cent, mainly thanks to the fact that the surveys took place during summer village festivities when local customs stipulate that the door of the house be open to any visitor. In Šumava, the response rate was 71 per cent.

The data obtained from the sample was compared with available census information from the two targeted areas (see Table 1.1). This confirmed that the survey sample included a large part of the local populations of the surveyed villages. However, it was hard to determine who was full-time resident rather than a second-home owner in the three Pelister villages, since the number of people who were registered as locals was not the same as the figure of inhabitants who lived in the area throughout the year. Some of the registered residents lived in the area for only one or two seasons, most often during the summer. There were also a number of people who were not registered as local residents (i.e. they were listed as permanent residents of the city of Bitola) but whom had decided to move to the Pelister area after retirement. Thus, the survey sample also included second-home owners and locally-born villagers whose official place of residence may have been elsewhere, but nevertheless owned a home in the park and were present at the time of the survey.

Table 1.1   Comparison between the number of survey respondents and local village populations

	Village	Number of households	Number of residents	Number of respondents in total	Local respondents	Second home owner respondents
Pelister	Nizhepole	25	87	54	22	32
	Brajchino	61	134	44	31	13
	Malovishta	38	98	33	23	10
Total		*124*	*319*	*131*	*76*	*55*
Šumava	Borová Lada	54	276	29	27	2
	Srní	128	343	52	39	13
	Horská Kvilda	15	73	15	11	4
	Prašily	47	153	31	27	4
	Modrava	21	52	17	12	5
	Kvilda	54	175	39	30	9
Total		*319*	*1072*	*183*	*146*	*37*

Sources: * State Statistical Office of the Republic of Macedonia, 2002 Census; ** Czech Statistical Office, 2001 Census.

The questionnaires used in the two parks were identical to each other, except for three questions that referred to local conditions and problems. Each questionnaire consisted of three blocks of questions. The first of these dealt with the socio-demographic characteristics of the respondents, the second focused on 'environment and nature in the national park', while the third section was related to tourism and sustainable development in the parks. The questionnaire structure and sampling approach utilized by the survey was based on the methodology that has been developed and implemented in the Czech Republic for approximately 10 years (Cihar and Stankova 2006). By using this approach I hoped to provide for a greater degree of comparability between my study and similar work undertaken in the Czech Republic and elsewhere in central Europe.

The data from the completed questionnaires was entered into database files employing MS Access and Excel. StatView was then used to analyse the data. In order to uncover the dependence among socio-demographic features and variables regarding the perception, attitudes and opinions of local residents towards the parks and their management, the chi-square (χ^2) test of homogeneity was used through cross-tabulation. This was applied to each park separately, and for the analyses of the results between the parks as well. The results that yielded $p = 0.05$ were considered borderline statistically significant. The significance level ($\alpha = 0.05$) was compared to the achieved p-value after the test was performed. In the cases where $p < \alpha$, H_0, the null hypothesis was rejected, demonstrating a relation between the two variables (Zvára 2003). During the statistical analyses the following requirements were considered: N (number of questionnaires) ≥ 30, expected frequency ≥ 1, and frequencies < 5 being no more frequent than 20 per cent (Zvára 2003). Additionally, the differences in the answers between the two national parks were assessed via a t-paired test. The consistency of the respondents' perception, opinions and attitudes was interrogated through the analyses of their answers in correlation tables.

Considering that one of the main aims of my research – as outlined in the introduction above – was to provide a more nuanced view of the local residents' understandings and experiences of nature protection in the two parks, I then set out to explore the different ways in which the surveyed residents' perceptions of the multiple dimensions of national park governance related to their residential attachment to the area in which they live. I analysed and cross-tabulated the evaluation of the statement 'I feel at home in the national park' (using a five-level Likert scale ranging between 'strongly agree' and 'strongly disagree') against those relating to the appraisal of the two national park authorities' local service management, nature protection, and educational and cultural functions. Consequently, I calculated the value of Pearson's *r* coefficient for each of the cross-tabulations.

Numerous survey respondents indicated that they would not mind being contacted again for an in-depth semi-structured interview. Out of this group, I approached a total of 25 local residents of the two parks, with the aim of obtaining a more nuanced insight into the complexities of local opinion, beyond the broad-level data provided by the survey. I also spoke with national and local policy makers, advocacy activists and business sector representatives (four in Pelister and three in Šumava). The interviews, which lasted approximately one hour, were recorded and later transcribed.[2] They took place during the summers of 2008 and 2013 (Šumava), and 2009 and 2012 (Pelister). The interviews were combined with information from official policy documents and locally-published literature to obtain a more comprehensive overview of nature management and protection issues in the two parks.

Alongside this work, the book also presents the results of a separate study of visitor-related issues in Pelister National Park. This piece of research is based on quantitative and qualitative evidence about the features and expectations of urban visitors to the park during August 2006. It involved interviews with 230 visitors at the main entry point in the park (see Figure 1.2), using a pre-set questionnaire. Interview questions focused on: i) the socio-demographic characteristics of the respondents; ii) the frequency of visits, length of stay in the park and the method of movement (i.e. by car, by foot); and iii) the visitors' needs, attitudes towards, and perceptions of, the protection and management of Pelister National Park. The data gathering process was supplemented by in-depth interviews with national park officials, visitors and local residents. In this book, the results of both sets of data have been compared to the outcomes of an earlier research project with similar aims, undertaken during the winter of 2003 under the auspices of the European Agency for Reconstruction.

Chapter Description

The remainder of this book is divided into seven chapters, which investigate the themes and questions highlighted above in further detail. The first three chapters provide a general overview of the theoretical and conceptual issues surrounding local participation in national park governance at the global scale, while chapters 4 to 7 are more directly focused on the CEE setting (with the Šumava and Pelister case studies being presented, respectively, in chapters 6 and 7). It should be pointed out that the two case study chapters are not analogous in structure and form, due to differences in their underlying evidence base, as well as the variegated challenges faced by each protected area. The empirical analyses that they present explore some of the key systemic problems faced by each park, rather than being developed along comparative lines.

Chapter 2 ('From Place Attachment to Wilderness: Key Concepts') contains a general discussion of the five main bodies of research considered in this monograph: national parks, communities, place attachment, rewilding, and tourism in protected areas. This review is necessitated by the broad and interdisciplinary nature of the

---

2   Some of the local resident interviews are cited in chapters 6 and 7. Interviewees' names have been changed in order to protect the respondents' identities.

knowledge corpus that I interrogated in order to construct my arguments and explore the available evidence base. One of the key claims advanced in the chapter is that a single understanding of all of these concepts does not exist; they are by definition contested and contingent.

Chapter 3 ('Protected Areas and National Parks at the Global Scale: A Critical Survey') provides a broad-level overview of the political and environmental contexts aiding the emergence of various nature protection practices throughout history. It explores the reasons for the rise of protected areas in developed- and developing-world countries, and the societal and biological roles served by nature conservation. This discussion is situated within a critical framework that highlights the multiple dissensions associated with protected nature across the world.

In Chapter 4 ('Obstacles, Victims and Opportunists: Conceptualizing the Relationship between Local People and Nature Protection') I explore the various ways in which the local residents of protected areas have been conceptualized in academic literatures and policy practice. I outline the political and historical underpinnings of three distinct approaches that have emerged successively during the twentieth century, and that see local people as either obstacles towards, victims of, or opportunists with regard to, nature protection. The chapter argues that the policy incarnations of all three frameworks exist simultaneously across different parts of the world, creating a hybrid geography of nature protection regimes with respect to the rights and roles of local people.

Chapter 5 focuses on the legacies, challenges and opportunities associated with nature protection and municipal government both during and after socialism. This part of the book highlights the manner in which environmental transitions in CEE are embedded in the broader transition from a centrally-planned to a market-based economy. The restructuring of environmental governance practices is explored in relation to the legacies of the centrally planned economy during communism, and development trajectories followed during the EU accession process. Particular attention is paid to the post-communist restructuring of nature protection practices and legislation, as well as the institutional processes of natural resource neoliberalization, and the multiple ways in which 'market' relations have entered the domain.

Chapters 6 and 7 provide in-depth accounts of the political, social, economic and spatial challenges faced by the governing authorities and local people in Šumava and Pelister, respectively. Developing further the ideas presented in chapters 2 and 3, this part of the book investigates the different ways in which local residents are attached to, and participate in, the governance of the national parks in which they live. I argue that local populations cannot be treated as a monolith in terms of their views of protected area management, which are themselves not necessarily correlated to perceptions of place attachment. In a broader sense, the chapters are aimed at evaluating the 'real life' implementation of the co-management model for protected areas. While many protected areas in CEE have been designated at the state level and are managed by hierarchically organized and rigidly-structured authorities – which are at the same time strictly controlled by the respective central governments – the chapters also explore how local communities often devise creative and diverse strategies beyond the victims-obstacles-opportunists triptych described in Chapter 3.

Park-specific issues are also investigated here. Chapter 6 focuses on the multiple local contestations of tourism development in Pelister, using both visitors and local people's perceptions as entry points. In Chapter 7, the 'bark beetle' controversy in the Czech Republic serves to illustrate the multiple ways in which rewilding strategies have been politically enrolled in national park management issues at the local scale. The conclusion summarizes main findings of the book, placing them in the context of some of the wider debates outlined in the introduction. It highlights, *inter alia*, the diverse and multi-sited economies associated with protected area management in CEE. The implications of the book's arguments for policy and theory are also outlined here.

Chapter 2
From Place Attachment to Wilderness:
Key Concepts

This chapter outlines some of the key concepts that constitute the theoretical basis for exploring the primary and secondary evidence in the chapters that follow. Using material sourced from a range of academic literatures, I focus on the multiple meanings and implications of scientific debates in five thematic areas relevant in this context: national parks, communities, place attachment, rewilding, and tourism in protected areas. In addition to offering a reference point for some of the conceptual and empirical explorations that take place in the rest of the book, one of the key purposes of this review is to illustrate in further detail some of the broader controversies arising at the local people-nature protection nexus.

There is a consistent line of argument running throughout the interrogations presented in the chapter: *relations that arise at the interface of society and nature are always contested and contingent.* As such, the discussion is invariably grounded in existing debates in human geography and political ecology, which have sought to highlight the complexity and ambivalence of ideas about nature (Castree 2013). As emphasized by Rytteri and Puhakka (2009), at the core of struggles over the establishment of national parks lie debates over the kinds of values that should be protected in nature. As a result, 'the conservation of nature's values answers a fundamental moral question about proper restrictions on the economic utilization of nature' (2009: 102). It is precisely the multiple contestations of the underlying reasons for nature protection – albeit in different thematic and geographical contexts – that provide the lynchpin for my exploration of the five conceptual domains reviewed in the chapter.

Defining Protected Areas and National Parks:
Conceptual and Practical Challenges

This book is largely premised upon the notion of 'protected areas'; Territories that are subject to a separate regime of environmental management due to containing recognized natural, ecological and/or cultural values. In the context of such spaces, the term 'protection' suggests that patterns and practices of human occupation, or the exploitation of resources, are somehow limited or regulated. As a result of the different spatial, cultural, economic and environmental roles played by protected areas, however, a situation has emerged whereby their sustainable management is now a major part of decision-making processes across the world (Antrop 2001).

This suggests that protected areas need to 'be considered as a landscape where trade-offs between nature protection and [the] socio-economic aspirations of local communities are expected to be well balanced' (Kušová et al. 2008: 38). The multiple purposes and meanings afforded by such territories are reflected in their most widely accepted international definition – provided by the International Union for the Conservation of Nature (IUCN) – where a protected area is seen as:

> A clearly defined geographical space, recognized, dedicated and managed, through legal or other effective means, to achieve the long-term conservation of nature with associated ecosystem services and cultural values. (Dudley 2008: 8)

Protected areas may often vary in terms of size, type and level of protection, depending on the enabling laws of their host countries, or the regulations of the international organizations involved in setting them up. In some cases of 'Marine Protected Areas', such territories may include parts of seas and oceans. Not all protected areas are public, either: The IUCN's definition of a Private Protected Area, as stated at the 2003 World Parks Congress, is:

> … a land parcel of any size that is 1) predominantly managed for biodiversity conservation; 2) protected with or without formal government recognition; and 3) is owned or otherwise secured by individuals, communities, corporations or non-governmental organisations. (Mitchell 2005: 1)

National parks are among the most widespread and well-known types of protected areas at the global scale. Their establishment dates back to the nineteenth century, and can be attributed to a range of social and political circumstances. In declarative terms, at least, the first national parks were created in order to safeguard pristine natural sites and to preserve the wild character of natural phenomena. Nature protection is still one of the dominant functions of such areas, although their meanings and purposes have evolved over time. The IUCN currently sees a 'national park' as:

> … large natural or near natural areas set aside to protect largescale ecological processes, along with the complement of species and ecosystems characteristic of the area, which also provide a foundation for environmentally and culturally compatible spiritual, scientific, educational, recreational and visitor opportunities. (Dudley 2008: 16)

This definition reflects the current policy emphasis on the protection of both natural and cultural assets. However, it is not applicable to all national parks at the global scale, due to a set of historical and spatial contingencies that have resulted in the application of this designation to areas with very diverse management systems and protection regimes; put simply, the existence of a common definition should not be taken to mean that a unified management system for such protected areas currently exists at the global scale.

The process of formulating standardized international guidelines and definitions in this domain dates back to the 1930s, where it received attention at events such as the London Convention in 1933 and the Western Hemisphere Convention in 1940 (Chape et al. 2008). Yet the formulation of a unified set of management criteria for national parks was only discussed at the First World Conference on National Parks in 1962, which recommended that the then International Commission on National Parks (today's World Commission on Protected Areas) 'establish a clarification of terms concerning national parks and equivalent reserves'.

Parallel developments within the IUCN – especially discussions held at the organization's General Assembly held in New Delhi in 1969, and the Second World Conference on National Parks in 1972 – ensured that the development of a common approach towards national park governance stayed high on the international agenda. It led to the IUCN's adoption, in 1978, of a classification system based on 10 nature protection categories; this was reduced to the current five following a review process that was undertaken between 1984 and 1990. Still, as the rest of this book makes clear, the meanings, practices and policies associated with the 'national park' concept remain contested and complex.

Recent decades have seen an increased awareness of the societal importance and role of national parks. This is a far cry from earlier policies and understandings of such areas, which emphasized the prevention – or elimination – of resource exploitation and even human residence, and did not include privately owned land. It is now widely acknowledged that there are many places where humans play a vital role in the landscape and ecosystem processes, and that inhabited places and systems are also in need of protection. Current mainstream understandings of this process emphasize that efforts to conserve nature must be part of a wider system of environmental management, encompassing different levels and modes of interaction with humans. However, the fact that such perspectives have historically failed to receive widespread acceptance – and are still marginal in some parts of the world – has become a major source of political tension. I explore such issues further in the chapters to come.

Variegated Understandings of Community

The case study evidence presented in this book, and the literature reviews in which it is embedded, are premised upon the notion that 'community is a fundamental modality for the conduct of modern politics' (Watts 2004: 195). Nevertheless, this has been a highly contested notion, to the extent that defining it is highly challenging and almost always forced to face a particular body of critique. Developing a conceptual framework based on the notion of 'community' in a rural setting that has encountered turbulent socio-ecological and political transformations makes this process even more complex. Still, instead of formulating yet another 'community' paradigm, I aim to open a discussion on how we can understand the complex networks that both shape and are shaped by processes that underpin the drawing together of people, place and meaning associated with this idea.

Interpretations and theorizations of the notion of community in the academic literature have mainly been developed within particular disciplinary boundaries. While sociologists have mainly understood 'community' as a social assemblage of spatially bounded groups, in anthropology the focus has been on clusters that are culturally linked (Delanty 2013). Conversely, philosophers have mainly predicated their discussions of community upon explorations of particular ideologies and utopian imaginaries.

As a discipline, geography mainly recognizes two types of communities: of 'place' and 'interest'. The former are defined via the borders of physically contiguous territories, which means that individuals belonging to a given community share a common space. For these types of communities, place is a proxy for face-to-face relationships. The heuristics of place attachment, place identity and sense of place – some of which are outlined in the section to follow – have largely been developed on the basis of this set of ideas (see Proshansky et al. 1983, Altman and Low 1992). The notion of territorial belonging, however, becomes much more stretched and relational when one starts speaking of communities of interest. Morrill (1987: 251) focuses on the 'common sense of identity and value across a range of interests' that are shared by people belonging to such groups. The idea of a 'transcendent community of interest' has also been proposed here, mainly in relation to 'the ties that bind across rivers, over mountains, skipping and jumping over the physicalities of geography' (Kelly 1995 cited in Leib 1998: 689). This concept has mainly been used as a marker for the evolutionary and fluid cultural identity of African Americans, which is partly a function of the fact that 'historical and contemporary racist policies and practices tend to replicate their effects in a variety of places, without regard to geography, and that people of color in similar circumstances grapple with similar problems' (ibid.).

Despite the divergent understandings of the concept within contemporary disciplinary boundaries, community has historically been theorized in the context of the relationship between politics, society and the state. The overall perception of community as a socio-political phenomenon has been based on its variable standing with and within society – i.e. whether it is the reflection or antithesis of society. The differential ability of community to act as a political subject has been seen as either advantageous or detrimental to its relationship with the state. Another common line of thought in more generic discussions of community refers to the loss and subsequent revival of community brought about by twentieth century developments. It has been argued that the vices of capitalist societies have killed off the notion of community, defined as a group of people who share values or beliefs, are mutually and emotionally related, and have frequent social interactions (Gläser 2001).

Many such debates can be traced back to the notions of *Gemeinschaft* and *Gesellschaft* (translated as 'community' and 'society', respectively), which were coined by the German sociologist and philosopher Ferdinand Tönnies in 1887 (Tönnies 2011). In his point of view, these two ideas represent the basic forms of human will. On the one hand, 'community' acts as a pre-modern (and often rural and familial) group, whose members are united by common beliefs and family ties, and whose will is organic and essential (ibid.). 'Society', on the other hand, is a modern group driven by practical, formal and impersonal relationships. Defined in this way, 'communities' tend to evolve into 'society'.

Understandings of community have also been developed in Émile Durkheim's work. Unlike Tönnies, he sees 'community' as a dynamic and flexible assemblage of human interactions rather than a rigid social structure bound by a fixed set of norms and relations (Durkheim 1973). Nevertheless, both Tönnies (2011) and Durkheim (1973) argue that intensive community relationships are only possible in pre-industrial, agrarian societies, in which people are socially and economically interdependent and bond by common traditional moral and religious conventions. Yet Durkheim (1973) does not believe in the return of the traditional and pre-modern community, but is hopeful about the creation of new communal bonds that will be instrumental to subduing conflict and disorder.

Not only did the question about the loss of valued traditional communal relationships dominate many research agendas in the social sciences throughout the twentieth century, but – as pointed out above – the disappearance of communal bonds was also perceived as the cause of some of industrial capitalist societies' key social problems. The rise of crime and illegality were seen as major consequences of a phenomenon termed 'social disorganization' by the Chicago School of sociologists such as Edwin Sutherland and Robert Lee Faris (Bursik 1988). Thus, the 1980s provided fertile ground for the re-emergence of 'communitarism': a term which was re-imagined from its original meaning in the nineteenth century (when it was formulated by John Goodwyn Barmby in 1841 in relation to utopian socialist societies) so as to emphasize the importance of common over individual and personal interests and rights (Etzioni 1995, Putnam 2000).

The downfall of communal connections and affects has also been related to the decline of 'social capital' – another highly contested concept. Although this idea was originally coined as far back as the early nineteenth century, it gained prominence only thanks to work by Jane Jacobs in 1961 (LeGates and Stout 2011), and was subsequently further developed by authors such as Bourdieu (1986), Coleman (1988) and Putnam (1995, 2000). In such work, social capital is often presented as the glue that creates and allows for the existence of strong and resilient social networks (Field 2003). In general, both Bourdieu (1986) and Coleman (1988) approach social capital as an individual asset, but the focus of their work is on different social groups. Field (2003: 28) criticizes Bourdieu's framework for being 'circular' and only tackling privileged groups who 'maintain their position by using their connections with other privileged people'. He also argues that although Coleman's thinking is more integrative and includes different types of individual, collective, disadvantaged and privileged actors, it is also:

> … naively optimistic; as a public good, social capital is almost entirely benign in
> its functions, providing for a set of norms and sanctions that allow individuals to
> cooperate for mutual advantage and with little or no "dark side". (Field 2003: 28)

Putnam (2000) is one of the most popular scholars working on issues of social
capital. His thoughts on the nature of civic renewal are mainly articulated in
relation to the manner in which the global spread of neoliberalism and the
withdrawal of the state from public life have shaped the character of developed-
world communities during the 1980s and 1990s (Rose 1999). To some extent, such
ideas have been echoed by Watts (2004) who argues that modern communities are
immersed in capitalism and market operations.

Social capital can be said to provide a systematic framework for analysing the
nature and function of contemporary communities. The different strengths – in
qualitative terms – of intra-communal relationships can be explained through the
lens of the constituent components of social capital, such as 'bonding', 'bridging'
and 'linking' (Kilpatrick et al. 2003), thus allowing for an improved understanding
of both the key social determinants of communities, as well as their flexibility,
resilience and adaptability (Adger 2010). Despite significant contribution to the
topic, accompanied by a broader attempt to expand the meaning and application
of the concept, social capital has been seen as a passive and receptive idea that
does not provide solid ground for political engagement and action, especially on
the part of excluded and disadvantaged communities (Putnam 2000). In response
to this situation, there have been calls for developing alternative theorizations of
communities, so as to embrace discussions on issues such as cultural fluidity, social
justice, and political dissensus (Swyngedouw 2007, Ranciere 2010, Escobar 2011).
Having noted that much of the work done on communities tends to present them
as a cogent, moral and social good 'to the exclusion of all else' Watts (2004: 197)
argues that:

> The unmaking and remaking of community, whatever their cultural or social
> content, and whatever their historical circumstances, must always address
> questions of representation (how they represent themselves and what forms of
> political representation they hold to), forms of rule, means of internal discipline
> and "purity", styles of imagination and their relation to accumulation and the
> economy. (Watts 2004: 198)

The portrayal of community as a unified and static idea has been strongly criticized
by postmodernists and feminists for its patriarchal undertones, and the tendency to
create new dichotomies, boundaries and exclusions (Young 1990). Watters (2003),
for example, has found that urban communities in the United States exist in the
form of 'urban tribes' that provide support networks and emotional ties beyond and
outside the conventional forms of social organization, thus dismissing Putnam's
claims as presenting an anachronistic and reductionist account of everyday life in
contemporary cities.

Scholars working in the field of nature conservation and resource management have been reluctant to theorize communities in an integrated manner. This is despite the fact that the essential components of conventional debates on the loss and revival of community have filtered through (either intentionally or unintentionally) to policy practices associated with the governance of protected areas. Indeed, it can be argued that the annihilation or collapse of a given community in a protected area stems from the application of fortress conservation approaches; these are instances where local communities have been displaced, impoverished and destroyed in the name of nature protection. Subsequently, collaborative community-based management approaches have helped strengthen the social bonds, physical conditions and spatial contingencies that allow such formations to exist.

Agrawal and Gibson (1999) are among the few scholars who have been brave enough to immerse themselves in efforts to theorize community in relation to resource governance. In their discussion of community, these authors (1999: 629) focus on three specific aspects: 'community as a small spatial unit, as a homogenous social structure, and as shared norms'. However, their conclusion leads to the recommendation that it may be more beneficial if community-based approaches can focus on institutions rather than communities. The idea of replacing communities with institutions has been further developed by Berkes (2004: 623) who claims that we should be cautious with the use of community, which he sees as a 'complex phenomenon' that hides a great deal of complexity in regard to conservation practices as a result of the fact that 'social systems are multiscale' and 'communities are elusive and constantly changing' (623). Developing further Brosius et al.'s (1998) argument that such formations are often seen in an idealized manner in public governance practices, he subsequently notes that 'as many conservationists know, it is often difficult to find a cohesive social group to work with in the field' (Berkes 2004: 623). Berkes (2004) thus suggests that is more useful not to focus on ever changing communities but on well defined, rules based institutions, which are able to cope with change in a structured way (Berkes et al. 2003).

However, the emphasis on institutions (rather than communities) is principally predicated upon the experience of implementing nature protection and natural resource governance programmes in a particular set of socio-ecological, cultural and spatial settings. As such, the proposition may not be applicable at a global scale. Perhaps Brosius et al. (1998) and Berkes' (2004) claims originate from the difficulties associated with a conceptualization of communities as inherently dynamic and contingent upon contemporary social and economic processes. The consequence of this set of circumstances has been a situation in which the relationship between community and nature protection has persistently failed to inspire new theoretical curiosity. The quandary can be potentially overcome, however, by further developing Adams and Hulme's (2001: 193) claim that community conservation 'is not a project or policy "choice" that can be simply accepted or rejected' (ibid.).

Nevertheless, it is clear that a rather outdated and romanticized notion of community – often based on Tönnies's notion of an organic and rural *Gemeinschaft* – has been universally employed for purposes of natural resource and protected area governance. It can also be argued that some of the unsuccessful outcomes of both fortress and community-based governance practices can be attributed to the inability to share context-appropriate knowledge, or theorize and understand community from a more holistic perspective, even though such programmes are meant to reflect precisely such a viewpoint. In part, this may be attributed to the fact that, while general community thinking has been mostly influenced by developments and empirical insights from Western developed societies, disappointments related to the application of community in conservation studies mostly comes from the experience of developing countries.

When speaking about the more specific empirical context of CEE, I would argue in favour of an empirical application of Watts' (2004) antinomies of community, whereby it is recognized that 'community-making can fail (often dramatically) because they may be unimagined or unimaginable' and that 'communities typically contain both reactionary and emancipatory tendencies'. Also relevant in this setting is the fact that the 'attendant forms of identity, rule and terrritorialization' allow communities to be 'produced simultaneously at different spatial levels (scale politics) and may work with and against one another in complex and contradictory ways' (198).

Unpacking Place Attachment

Moving onto the linkages that allow communities to articulate a particular relationship to the places in which they live, it can be argued that experience from the implementation of contemporary nature protection approaches across the world indicates that successful local participation cannot take place without genuinely taking into account the aspirations and needs of affected populations. Socio-economic status and family ties in particular have been shown to influence the manner in which people perceive nature conservation and the operation of park authorities, as well as the degree of involvement in processes of planning and management (Lane 2001). Less is known, however, about how these and other features of local people relate to the implementation of various protected area models, particularly with respect to the reconciliation of social and environmental goals (Kellert et al. 2000, Adams 2005).

In this context, it is worth mentioning the complex and rich literature on place attachment and home, which has elaborated a wide array of inter-related concepts, including, *inter alia*, 'community attachment', 'sense of community', 'place dependency', 'place identity' and 'sense of place' (de Sans 2004). This body of work has often explored the extent to which attitudes towards place, space and the management of national parks vary among different groups of residents and areas. Experts working on the subject have emphasized that 'people's senses of themselves are related to and produced through lived and imaginative experiences of home' (Blunt and Dowling 2006: 24). Such theorizations are grounded upon the notion, pioneered by Hummon (1990) that cultural images of places may be 'appropriated by individuals to elaborate self-conception', as for example, forms of dwelling and settlement types may 'provide residents with a community identity' (Cuba and Hummon 1993: 113–114, also see Hummon 1990).

Place attachment – which can be seen as the affective link that people establish with specific settings, where they tend to remain and where 'they feel comfortable and secure' (Hidalgo and Hernández 2001: 274) – is a key factor in this context. One of the most comprehensive theorizations of its multiple social and political dimensions has been provided by Altman and Low (1992), who note the diverse ways in which people-place bonds are articulated within different social and spatial scales. Explorations of the success of environmental management strategies in different geographical settings have also used place attachment as an explanatory framework. Thus, Vorkinn and Riese (2001) note that the development of close socio-economic and cultural linkages to an area results in a higher degree of sensitivity to site management and impacts among local communities. At the same time, Kaltenborn's (1998) study of the effects of sense of place on responses to environmental impacts in Svalbard claims that people's reactions to environmental impacts vary according to place meanings and affiliations. Operating within the context of protected areas, Payton et al.'s (2005) work has argued that place attachment is a significant predictor of civic action in a wildlife refuge, while Smaldone et al. (2005) have found that Grand Teton National Park contains a multiplicity of place meanings, many of which are overshadowed by more easily visible scenic or recreation opportunities.

Working in CEE, Lawrence (2008) has found that national parks are characterized by conflicts between centralized decision making and new public values and concerns. She emphasizes the need for new participatory approaches, since post-socialist countries lack the civil society and resource management structures that characterize societies with established nature conservation models (also see Grujičić et al. 2008, Rodela and Udovč 2008 and Švajda 2008). In terms of place attachment issues, Castán Broto et al.'s (2010) interrogation of the relationship between identity and place among residents living around five coal ash disposal sites in the Bosnian city of Tuzla has found that bonds between people and places in environmentally degraded settings are 'not necessarily dislocated', and may even be strengthened by individuals' performances of adaptability and resilience in such contexts (952). Their research has highlighted the strong relationship between perceptions of nature protection, on the one hand, and the character and strength of place attachment among local residents, on the other.

One of the overarching conclusions that can be drawn from this body of work is that local understandings of the spatial aspects of nature conservation play a central role in determining both the character of protected areas, as well as the success of environmental management policies in them. The suggestion, therefore, that national parks are deeply embedded in local historical and geographic circumstances is one of the key premises of this book.

'Wilderness' and its Discontents

Besides providing one of the core tenets for the foundation of national parks in the nineteenth century, the notion of 'preserving' and 'bringing back wilderness' is also central to the Šumava case study that is presented in Chapter 7. Yet 'wilderness' is not necessarily the first image that comes to mind when thinking of Europe and the European landscape. Not only has this continent traditionally been perceived as a densely populated area whose natural features have been deeply affected by human civilization, but it is also the case that European metropolitan areas served as a counterpoint to the 'wilderness' concept in North America (Lupp et al. 2011). This is illustrated by the experience of Aldo Leopold – one of the founders of the 'wilderness' protection movement in the United States – who, during his visit to Europe in 1935, was disappointed that wilderness had vanished from the continent's surface, as well as human memory (Meyer et al. 2009).

The last three decades, however, have seen the emergence of a 'rewilding' agenda in Europe – expressed in the form of policies aimed at returning a given area to an imagined 'wild' state – as a result of the influence of a complex array of mutually intertwined social, political and natural factors. If Europe partly inspired natural scientists and practitioners to protect wilderness areas in North America, it can be argued that the same wilderness, especially that of national parks such as Yellowstone and Yosemite, helped stimulate the idea of human-led rewilding in Europe (Lupp et al. 2011).

Anthropogenic rewilding efforts first occurred in Europe at the beginning of the 1980s. One of the first initiatives of this kind – 'Trees for Life' in northern Scotland – focused on the restoration of 600 mi^2 (1,554 km^2) of the ancient Caledonian Forest in 1984. This was followed by the establishment of similar areas across the continent: some of the largest ones include Val Grande National Park – 'the largest wilderness area of the Alps and Italy' (Höchtl et al. 2005: 85) and the 'Wild Heart of Europe' – a joint wilderness area shared by Šumava National Park and the Bavarian Forest National Park in Germany (discussed further in chapters 3 and 7). Wilderness areas are often synonyms for the 'core' zones of the national parks, where human impacts have been historically minimal, or management activities have been halted (Diemer et al. 2003).

Although the designation of wilderness areas in Europe started in the 1980s, their meaning was not associated with a formal definition for a long time. At the national level, only Finland defined the establishment of a protected area in this domain as a special category in the Act on Wilderness (Gladden 2001). In the EU's Guidelines on Wilderness in Natura 2000 (European Commission 2013), a 'wilderness area' is defined as:

> … an area governed by natural processes. It is composed of native habitats and species, and large enough for the effective ecological functioning of natural processes. It is unmodified or only slightly modified and without intrusive or extractive human activity, settlements, infrastructure or visual disturbance. (2013: 10)

Respectively, according to these Guidelines, *wild areas* are:

> … often smaller than wilderness areas. Here the original natural ecological conditions have been slightly modified by extractive activities such as forestry or other extensive human activities. These fragmented areas can support natural processes typical for larger areas if they are connected through functional ecological corridors to the surroundings. Wild areas sometimes have the potential to become wilderness by the process of restoration. (2013: 76–77)

The document, therefore, introduces a normative element to the rewilding process, by distinguishing between areas that have already achieved a given ideal, and those that should reach it via concerted policies. As such, it reflects a growing academic argument that Europe is a strong candidate for rewilding policies, thanks not only to the existence of contiguous areas with relatively little disturbance but also due to the relatively recent date of megafauna extinction on this continent.

Rewilding efforts have also been boosted by unrelated background developments, such as the intensifying abandonment of arable land, and the creation of fallowed areas. Farming in marginal regions has been rendered economically unfeasible by structural changes in Europe's common agricultural policy, and a decrease in farming subsidies (Tickle 2000, Theil 2005). Mid-range estimates of the Institute for European Environmental Policy (IEEP) have shown that approximately 168,000 km^2 of farmland might be abandoned by 2030 in the EU (Keenleyside and Tucker 2010). Some experts have seen this process of 'natural rewilding' as a unique opportunity for the establishment of new 'wilderness areas' in Europe (Zunino 1980, Theil 2005, Martin et al. 2008).

Further supporting the 'rewilding' agenda have been the uncertainties brought about by impacts of climate change on various socio-natural systems, as well as the constant loss of biodiversity. As was noted above, the fall of the Iron Curtain allowed access to sizeable, intact areas of relatively low anthropogenic influence in CEE, with a significant prospect for experimental 'rewilding' (Lupp et al. 2011). Another factor is the increase in NGO activism on the issue (Meyer et al. 2009), as demonstrated by recent initiatives undertaken by the Wilderness Foundation (based in both Germany and the UK), the PAN Parks Foundation, IUCN, WWF and Europarc Federation. The protection and restoration of wilderness areas in Europe has also been promoted by the Green Belt programme. Wild Europe and Rewilding Europe have also been involved in this initiative: while the former brings together large international organizations such as the WWF, UNESCO and the Council of Europe, the latter includes WWF-Netherlands, Wild Wonders of Europe and ARK

Nature. It is worth noting that Rewilding Europe recently received a grant of 3.1 million Euro in order to rewild 1 million ha in five abandoned agricultural areas in Europe – mostly CEE – by 2020.

Even though rewilding efforts are still nascent and relatively fragmented in the European context, they have flavoured many recent debates on national park management. In this book, rewilding provides the context not only for the Šumava case study, but for broader discussions concerning the reflections of different levels of human intervention in protected areas upon local democracy and public participation.

Visitor-related Issues in Protected Areas

The conditions and practices associated with the management of national park visitors and tourists are one of the key determinants of the functioning of these protected areas; part of the Pelister case study presented in this book explores the everyday grain of such questions. The importance of visitors to protected areas can be traced back to the romantic strive to 'modernize' cities during the industrial revolutions that engulfed western Europe in the nineteenth century, with the aim of offering an alternative to the smokestacks of the industrializing economy (Suckall et al. 2009). In abandoning the principles of science and reason, romantics stressed the power of imagination, feeling and emotion. They focused their attention on the aesthetics of the natural world, transforming places that had once been seen as 'valueless' into 'picturesque and sublime' natural refuges (ibid. 1196). Over time, however, there is ample evidence to suggest that national parks gradually moved from acting as the providers of a romantic 'escape' from the contamination of the city, to active players in the formation of the identities and lifestyles of their visitors.

However, this trend had an important class dimension, as the UK's national parks, for example, were generally placed in regions that affluent, university-educated white people deemed important (Suckall et al. 2009). This means that their aesthetics were perceived and designed by individuals who had been educated in the romantic idea of 'wilderness and solitude' (ibid. 1197). The upper class romantics were keen to keep the landscape exclusive to those who had the 'financial and cognitive' resources to appreciate 'scenery, landscape, image, [and] fresh air' (Williams 1972: 6). It can be argued, therefore, that national parks are often managed to preserve tranquillity and 'naturalness', even if this results in a contrived landscape. Suckall et al. (2009) point out that such an approach leaves little room for dynamism and evolution, since any new transformations may be seen as less aesthetically desirable by national park visitors.

The fact that urban dwellers' preferences and needs regarding natural sites have varied through time and space has also been picked up by authors such as Beunen et al. (2008) and Matsuoka and Kaplan (2008); in order to determine the changing character of contemporary urban perceptions of nature, the latter have undertaken a comprehensive analysis of 58 volumes of the journal *Landscape and Urban Planning*, discovering that much of the surveyed work addresses nature as a place that can help improve the quality of life and provide a brief sanctuary from 'urban problems'.

However, more recently there has been increasing academic interest in the role of visits to protected nature as a signifying symbol of status, fashion and a healthy lifestyle. The main question in this context pertains to the tension between urban dwellers' desire to explore and protect nature, on the one hand, as opposed to their interest in entertainment and relaxation, on the other. Considering that the latter can be provided even at the gateways of national parks, Beunen et al. (2008) have stressed that most urban dwellers rarely venture into the cores of protected areas, choosing to use the gateways for recreational activities. Similarly, Sterl et al.'s (2008) assessment of the awareness of the disturbance of wildlife due to different anthropogenic uses – including recreational activities – among on-site visitors in Donau-Auen National Park, Austria, showed that only 12 per cent of the respondents believed that they could have potentially disturbed wildlife on the day of the interview.

The mass endorsement of an active and healthy lifestyle, which helps reaffirm the connection between individuals and nature (Eagles and McCool 2002) has combined with the expansion of low-cost airlines to make nature conservation areas and activities around the world more easily visible and accessible, thus further accelerating the development of nature-based tourism. The effects of such developments are particularly pronounced in national parks, as, according to Obua and Harding (1996), people are generally more attracted to a particular area if they know that it has been designated for protection, since this affects the perception of its attributes and natural beauty. Even though the ideology behind nature-based tourism is based on nature protection principles, its environmental consequences can potentially be more damaging than mass tourism, primarily because they are mainly concentrated in sensitive and vulnerable areas (Gössling 1999). The increasing interest in nature-based tourism in protected areas, especially those located in developing countries, is also magnifying and diversifying its implications for the development of local communities (Goodwin 1996, Lindberg and Johnson 1997). This means that the impacts of tourism activities in this domain extend beyond the natural environment (Goodwin 1996), since they also affect socio-economic processes and municipal governance.

It should be pointed out that a linear relationship between visitors and local communities does not exist. According to Eagles and McCool (2002: 27), interactions between visitors and local communities can include cultural (exchange of folkloric and traditional habits, customs and beliefs), economic (investment, boost to the local economy, changing labour markets), demographic (changes in population structure, migration processes), environmental (pollution, destruction of natural sites, natural habitat disturbances) and political (changing laws, designation of more protected areas as a result of increased awareness and appreciation for the parks) dimensions. That is one of the main reasons why Mehta and Kellert (1998) argue in favour of integrated and simultaneous planning of local development, including tourism and nature protection. Indeed, as Obua and Harding (1996: 505) point out, 'understanding visitor characteristics is a principal aspect of sustainable tourism' and 'planning of new developments, management of visitors, monitoring trends and predicting recreation demands all require adequate information on visitors'.

In this context, it is worth mentioning Bourdieu's (1986) maxim that a person's 'lifestyle' is a trade-off among preferences relating to leisure, cultural consumption, and cultural tastes. Thus, many authors stress the need for establishing the profile of national park visitors within the context of their affiliation to a particular social group, and in terms of their individual preferences for different cultural and leisure activities (Featherstone 1991, Suckall et al. 2009). According to Breakell (2002) most of the national park visitors in Britain and western Europe are from a middle class background; a finding that is also supported by, *inter alia*, Agyeman (1995) and Agyeman and Spooner (1997). Similarly, Obua and Harding's (1996) work in Uganda has indicated that most visitors (52 per cent) to Kibale National Park are university graduates. Petrosillo et al. (2007) discovered that marine protected areas in Italy had a higher rate of male visitors, whose overall sample was otherwise dominated by the 31–45 age group, and individuals with a secondary education.

The formulation of a visitor management strategy within the comprehensive management plans of protected areas must therefore be predicated on an adequate understanding of visitors' preferences and perceptions. Therefore, Petrosillo et al. (2007) argue that socio-economic status, cultural ties, and past experiences are key factors influencing the perception of environmental quality. The preferences of urban dwellers towards activities in national parks can differ in many ways, based on their personal characteristics, education levels and the place of residence (ibid.). For example, Matsuoka and Kaplan (2008) have underlined that younger people are attracted to more active pursuits (i.e. sports activities), while adults and the elderly would rather enjoy nature opportunities that afford observation and relaxation (Oguz 2000, Sherman et al. 2005, Oku and Fukamachi 2006). Moreover, it is thought that urban dwellers show a greater interest in enigmatic and exotic environments (Herzog 1989, Ulrich et al. 1991, Hartig 1993). In light of the diversity of the visitors' needs and expectations, many authors thus believe that measures to influence tourist behaviour should be principally aimed at preserving the balance between nature and recreation (Reynolds and Elson 1996, Cope et al. 1999, Beunen et al. 2008).

Despite the existence of a significant body of knowledge on the issue, there is still an empirical gap regarding urban dwellers' perceptions of, and attitudes towards, national park management and protection. Basically their motivations for visiting such areas remain unclear and inadequately connected to contemporary social science research regarding the relationship between nature conservation and tourism. There is also a need for additional inquiry into the urban dwellers' influence on the parks, as well as their interaction with local communities. The necessity for understanding this relationship is particularly pronounced in central and eastern European countries, where many national parks still lack the monitoring capacity – and consequently the necessary political strategies – for visitor management, despite the extensive structural changes experienced by their host societies over the past 20 years.

Moving Forward

In providing a review of the five thematic areas listed above, I have aimed to demonstrate the benefits of thinking in an integrated manner – and across disciplines – in exploring nature protection-related challenges. The literatures outlined here have rarely entered into a dialogue with each other, resulting in a disjointed landscape of knowledge where issues of local participation, nature conservation, wilderness-orientated strategies, and the implications of tourism are often treated within separate academic silos. At the same time, and despite their presence in both exclusive and participatory programmes, the existence of communities of place and interest in protected areas has received inadequate theoretical attention by academics or practitioners working in the field. This means that theoretical understandings of the local people-protected area relationships have subsequently translated into policies that are often disparate and incoherent.

The five thematic areas will be explored further in the chapters that follow, albeit to a different extent and via different means. While tourism and wilderness represent, respectively, crucial concerns in the Pelister and Šumava case studies, questions of 'community' and 'place attachment' are present throughout the evidence base. The challenges associated with the definition and establishment of protected areas are principally explored in Chapter 3, although they basically permeate the entire book. Underpinning the interaction of this domain with other debates is a desire to understand 'how the various benefits and costs of nature conservation should be distributed' (Rytteri and Puhakka 2009: 103). This allows the management of national parks to be moved away from 'values' onto explorations of 'political interests and power relations' (ibid.) as well as the economic practices that underpin everyday life.

Chapter 3

Protected Areas and National Parks at the Global Scale: A Critical Survey

This chapter provides a broad-level overview of the political and environmental contexts associated with the emergence of various nature conservation practices throughout history. Particular attention is paid to the rise of national parks, as areas subject to particular nature protection regimes. The chapter explores the reasons for the designation and operation of national parks in different socio-ecological and spatial settings, as well as the broader political and biological roles played by nature protection. This analysis is situated within a critical framework that highlights the multiple dissensions associated with the diverse expert and policy-maker responses to the changing realities in which national parks exist.

My principal argument in the chapter is that national parks in reality represent a mélange of disparately managed areas, designated under specific historical conditions. As such, they are subject to a constant process of spatial evolution and transformation, in line with the changing importance and nature of the environmental, social and political services that they provide. Thus, national parks are not 'natural' areas; rather, they are entirely socially constructed, and constantly shaped by geopolitical, cultural and economic technical activities (Lefebvre 2001).

The chapter consists of two main sections, the first of which explores the historical setting and key driving forces for the protection of geographical places and biological species across the world. An investigation of the peculiarities of the demand for, and designation and management of, national parks, provides the basis for the second part of the chapter. In order to highlight the diverse regulatory and political reconfigurations of such areas through time and space, I survey the reasons for, and components of, the establishment and governance of such areas across i) The United States and Canada; ii) Latin America; iii) Africa; iv) Australia and New Zealand; v) Asia; and vi) Europe. In each geographical realm, I identify the distinctive features and practices that have shaped the functioning of national parks over time. The conclusion of the chapter critically revisits the multiple theoretical, discursive and decision-making incongruences that have contributed to the emergence of national parks as a whole.

The Rationale Behind Protecting Nature: A Brief Historical Overview

It would be impossible to imagine the development of human civilization without the extensive and persistent use of natural resources. Recognizing this fact, societies have made various efforts to protect the sites and areas that they have perceived as valuable in hosting such 'natural capital'. For example, the Mauryan kings of northern India designated nature reserves in order to conserve particular areas (i.e. forests) and species (such as elephants and fish) as early as 300 BC (Grove 1995, Chape et al. 2005). Initial efforts to protect nature also involved a strong cultural dimension, often being aimed at safeguarding various sacral artefacts and locales.

The distinction between private and public interests was highly pronounced in pre-modern nature protection initiatives. While specific flora and fauna were protected for broader public benefits, it was often the case that particular resources and areas were kept exclusively for the use of elites. This was especially pronounced in Europe, where feudal landlords and royal families maintained hunting grounds and forests for their own exploitation. Thus, the Bohemian Prince Konrad Ota issued the Conrad decree around 1189 and the Book of Rožmberk in 1360, for protecting forests from poachers. In the seventeenth and eighteenth century French nobility, artists and philosophers such as Louis XV, Jean-Jacques Rousseau and Voltaire 'rediscovered' their attachment to the countryside, alongside a specific sensibility towards 'picturesque' landscapes (Matagne 1998). Their enthusiasm opened up a vigorous debate regarding the relationship between humans and nature, as well as the perceived separation of the 'natural' from the 'cultural' landscape. Further east, Prussian rulers also made significant efforts to protect their forests during the eighteenth and nineteenth centuries (Zeide 2001).

The modern concept of nature protection emerged in the nineteenth century, under specific socio-political and historical conditions. Some authors (such as Jepson and Whittaker 2002) argue that two particular passions of the nineteenth century's elite societies from Europe and North America – natural history and hunting – were instrumental in the reconceptualization and re-perception of nature. This was the time of the European colonization of North America, Africa and Asia, supported by the construction of vast railway networks that provided the Old World aristocracy with new opportunities for travel and trade. The new infrastructures evoked the colonizers' curiosity for the exploration of exotic ecosystems in the new colonies, arousing a passion for 'unknown' natural beauty, and a special interest in natural history. The newly-occupied spaces were also abundant with large game, thus providing an attractive resource for satisfying the hunting passions of European elites.

Wildlife protection in the colonies, North America and Europe was mostly a result of the efforts of exclusive social clubs and lobby groups, which were created by influential individuals in London, Paris, Amsterdam, Brussels and New York by the 1880s (Jepson and Whittaker 2002). During this period, one of the most prominent organizations campaigning for wildlife conservation – as well as hunting in 'fair chase' – was the exclusive Boone and Crockett Club, created by Theodore Roosevelt in 1887. Members of the Club, who included some of North America's most eminent citizens, argued that wealthy and powerful individuals are more effective at protecting natural assets than bureaucratic governments. Contrasting their approach was Hugo Conwentz's *Naturdenkmal* model, which promoted the development of an integrated framework to bring together the protection of natural and cultural capital. This paradigm postulated that the relationship between humans and nature needs to incorporate i) the 'memorial' (*Denkmal*), allowing the commemoration of persons, arts and nature; ii) efforts aimed at preventing the destruction of nature; and iii) the emphasis on patriotic values in the preservation and nourishment of 'national' natural capital (Lekan 2009). *Naturdenkmal* societies were founded in France (1901), Switzerland (1909) and Britain (1912): their members promoted pragmatic resource management through the inventorization and protection of specific natural assets. In addition to being strongly supported by the middle classes across Europe, the *Naturdenkmal* concept also became official government policy in countries such as Sweden and Prussia (Ladle and Whittaker 2011).

In 1903, just a few years before the creation of the British *Naturdenkmal* society, the adherents of the Boone and Crockett Club in the United Kingdom established the Society for the Preservation of Fauna in the Empire (later known as Fauna Preservation Society). Considering that the majority of the Society's members originated from the governing elites, it comes as little surprise that their main emphasis was on government-led nature protection. While the *Naturdenkmal* societies mostly promoted nature protection at the level of the nation state in continental Europe, the Boone and Crockett Club and the Society for the Preservation of Fauna in the Empire were more concerned about the safeguarding of wildlife in the colonies.

Encounters with 'pristine' nature helped create an awareness of the negative environmental impacts of industrialization among influential individuals. The perception that European landscapes are already severely damaged and densely populated was a major source of anxiety about the possible loss of recently discovered natural resources (Matagne 1998). In addition, the sheer volume of hunting parties from Europe and North America resulted in excessive rates of extermination of game in the colonies, spawning the first ideas about the protection of natural places and species in colonial lands. As a result of such concerns, colonial governments were urged to place restrictions on hunting, while creating game reserves to protect wildlife.

However, the two main colonial powers in Africa at the time – Britain and Germany – had different ideas about the ways in which local wildlife should be protected. While the British believed that the game-law system of 'closed-seasons, licenses and schedules of protected animals' would be sufficient, the Germans were more favourably inclined towards the fenced game reserve concept, where hunting is limited or forbidden at all times (Jepson and Whittaker 2002). Although this debate remained open until the first half of the twentieth century, both governments established pioneer game reserves in the African colonies during the last decade of the eighteenth and the beginning of the nineteenth centuries (Waithaka 2012). The reserves were created outside the settlers' plantations and 'wherever wildlife was in conflict with farming, it stood no chance' (ibid. 25).

During the last decades of the nineteenth century, wilderness conservation campaigns were also very effective in North America, leading to the creation of the first modern protected areas. Yosemite was designated by the United States Congress as a pioneering example of a new national-level model of nature safeguarding. The first protected area to be named a 'national park' – Yellowstone – was established only a few years later, in 1872. During the years that followed, a few more national parks were created in the country – most of them covering vast, scenic and pristine areas in the western regions. They included Mount Rainier National Park (1899) and Sequoia National Park (1890).

While the national park concept spread swiftly across North America, Australia, New Zealand and Europe, wildlife conservation practices were still not consolidated across Africa. In 1900, this prompted the United Kingdom to organize an international conference in London, featuring representatives of all European colonizing powers: France, Portugal, Spain, Belgium, Germany and Italy. The main aim of the event was to 'facilitate the creation of uniform game regulations and law enforcement procedures within the African continent' (Waithaka 2012: 25). One of the most important decisions at the convention was the declaration of the killing of wildlife in the game reserves as 'unlawful' with the exception of animals characterized as 'vermin'. This included lions, leopards, hyenas, wild dogs, otters, baboons, some monkeys, large birds of prey, crocodiles, poisonous snakes and pythons (Sorrenson 1968). Such so-called 'vermin' were meant to be eliminated both within and outside the protected areas, predictably leading to detrimental effects on the overall extent of wildlife in the colonies (Western and Waithaka 2005).

This and many other similar policies have prompted the academic mainstream to argue that nature protection in the nineteenth and the beginning of the twentieth century was a controversial mixture of romantic feelings, scientific research, and political-economic interests. Matagne's (1998) comment on the nature protection policies from that period provides a good summary of such thinking:

> Everything concerning the domain of nature conservation is rich in paradoxes and contradictory interests: the romantic ideal; the established scientific facts; economical and colonial interests; patriotic feelings; moral, ethical and aesthetic arguments. (359).

One of the most controversial initiatives for 'protecting' wildlife was the Roosevelt-Smithsonian Institution expedition to present-day Kenya and Uganda in 1909 and 1910, led by the Boone and Crockett Club and the Fauna Preservation Society. The expedition resulted in the killing of approximately 13,000 wildlife under the justification that it might be too late for saving Africa's wildlife from extinction, as a result of the combined threats of rapid human expansion, poaching and intensive farming. It was claimed that such a wholesale killing of native fauna would provide the last opportunity to create specimen collections in museums.

Table 3.1   A global chronology of selected nature protection initiatives relevant to the establishment of national parks

Time	Process/event
10,000 BC	As agriculture began to transform the relationship between people and nature, local communities recognized specific sites as 'sacred', and protected them from certain human uses. Applied differently in different places over the subsequent millennia, the concept was a widespread practical measure that people found beneficial in both material and spiritual ways.
252 BC	Emperor Ashoka of India established protected areas for mammals, birds, fish, and forests: The earliest recorded areas where a government protected certain resources.
684 AD	The first Indonesian nature reserve was established by order of the King of Svinjaya, on the island of Sumatra. Sumatra is now recognized as one of the world's centres of mega diversity, with numerous protected areas – the major sites comprising the recently declared 25,000 km² Tropical Rainforest of Sumatra World Heritage site.
1079	William the Conqueror claimed the New Forest (England) as a royal hunting reserve and protected it against illegal harvesting from rural people; poaching became a major law enforcement issue, but timber from the forest was essential to England's war efforts in the seventeenth to nineteenth centuries. Today the New Forest is still a valued protected area; it became the UK's newest national park in 2005.
1800 onwards	Passion for natural history among European aristocracy. Founding of natural history museums.
1880–1900	Growing concerns over the impact of industrialization and clear felling led to efforts to create site inventories and habitat maps. First published for France in 1896.
1865	Yosemite (California) was established by US Congress as effectively the first of a new national-level model of protected areas; Yellowstone was the first area to be called a national park.
1900	The first Conference on African Wildlife took place in London in 1900. The resulting convention set up game laws as the main instrument of control, to be supported by the establishment of game reserves covering tracts of land of sufficient size to facilitate large-scale migrations.
1909	The first European forum for action was the 1909 Paris Congress for Landscape Conservation.
1925	The King Albert Park was established in Congo, as the first national park in Africa.
1933	The landmark London Conference on African Wildlife took place. This resulted in the world's first major convention concerned with wildlife preservation. The conference brought together 60 influential delegates from colonial powers with territories in Africa. The conference was held in the House of Lords and had royal patronage.
1978	The IUCN's system of categories of protected areas was published; it set the framework for a worldwide assessment of protected area coverage. Its latest revision took place in 1991 and is now being promoted for other management applications.
1980	The World Conservation Strategy – published by IUCN, WWF, and UNEP – popularized the concept of sustainable development and set up a partnership between conservation and development.
1981	The protected Areas Data Unit was established at the IUCN Conservation Monitoring Centre by the IUCN and its Commission on National Parks and Protected Areas; this provided the first worldwide database on protected areas.

Time	Process/event
1982	The Third World National Parks Congress took place in Bali. Indonesia emphasized the importance of protected areas as a key element in national development plans; it set a 10 per cent protected area coverage of each of the world's biomes as a target.
1987	'Our Common Future' (the Brundtland Report) by the UN Commission on Sustainable Development called for 12 per cent of the land to be given protected area status, and advocated global action to conserve biodiversity.
2002	The World Summit on Sustainable Development, Johannesburg, South Africa, called for the loss of biodiversity to be reversed by 2010, and for a comprehensive system of marine protected areas to be established by 2012.

Source: Adapted from McNeely 1994, Jepson and Whittaker 2002, Chape et al. 2008.

Participants in the Roosevelt-Smithsonian expedition in Africa believed that many local wildlife species were facing extinction threats, and that their long-term survival could only be achieved by isolation from humans. Using his understanding of the 'national park' concept in North America, Theodore Roosevelt – one of the key figures in the expedition – proposed the creation of nine national parks based on the 'fence' or 'fortress' model. The national parks that he suggested were conceived as large fenced areas that should provide sanctuaries for endangered wildlife. Such spaces were effectively meant to incorporate already existing game reserves, while achieving specific conservation objectives. However, deep divisions and disagreement among the attitudes and policies of colonial governments towards wildlife conservation in Africa persistently delayed the formation of national parks on this continent. The first such area – Albert National Park (today known as the Virunga National Park) – was only created in 1925. A similar situation transpired in Asia, where the Angkor Wat National Park was established in Cambodia in the same year.

The 'National Park' Concept Expands Further

While Africa and Asia had to wait a few decades for the designation of their first 'modern' national parks, Australia, New Zealand, South America and even Europe promptly accepted the concept immediately after the creation of Yellowstone. Thus, Australia's first national park– the Royal National Park – was created in 1879, followed by the founding of many other parks in the last decades of the nineteenth and the beginning of the twentieth centuries. Tongariro National Park was New Zealand's pioneering protected area of this type, having been instituted as early as 1894. Also significant in this context were the designation of El Chico National Park in Mexico and Banff National Park in Canada, which took place, respectively, in 1882 and 1898.

Most of the national parks created at the turn of the twentieth century were located in sparsely populated, economically marginalized and physically remote areas, which were seen as potentially valuable for tourism development and scientific research (Brovko and Fomina 2008). European conservationists were struggling to find such regions in the densely populated and industrialized spatial fabric of the continent. Still, the creation of the Sarek National Park in Sweden (1909) and the Swiss National Park in Switzerland (1914) demonstrated that these areas could also exist in Europe, marking the beginning of an extensive national park tradition.

The decades that followed saw an upsurge of national parks across the world. Although the general postulates of the Yellowstone concept were applied at a global scale, its practical implementation reflected various 'local' understandings, practices and relations. The United States conceived national parks as pieces of 'public land to which public entry for recreation and observation was facilitated, but wherein fauna and flora were preserved in a near natural state' (Jepson and Whittaker 2002: 142). As pointed out by Chape et al. (2008), in North America, the protected areas were meant to safeguard dramatic and sublime scenery. While the implementation of this concept in British colonies followed a similar approach, the French understanding – deeply influenced by a fascination with landscape and scenery – perceived such areas through a much more romantic lens. However, French, Belgian and Portuguese colonial governments regarded free public access as a threat to the conservation of natural landscapes.

The divergence of national views about the purpose and character of national parks was reflected in the lack of a consensual approach towards the management of national parks in African and Asian colonies. One of the main cleavages existed among supporters of the governance model led by 'committees of prominent individuals', on the one hand, and believers in the effectiveness of more formal governmental-style committees, on the other (Jepson and Whittaker 2002)). The growing effort to define an international common ground on the management practices underpinning protected areas led to the organization of an International Conference for the Protection of Fauna and Flora in London in 1933. The conference resulted in an international agreement that established protected areas as the fundamental mechanism for achieving wildlife conservation. It also aided the formulation of four protected area categories: i) national park; ii) strict nature reserve; iii) fauna and flora reserve; and iv) reserve with prohibitions for hunting and collecting (Phillips 2004).

The 'national park' category was also recognized within a separate four-pronged classification – used by the Pan American Convention on Nature Protection and Wildlife Preservation in the Western Hemisphere in 1942 – that also included national reserves, nature monuments and strict wilderness reserves. The existence of multiple parallel systems for the categorization of protected areas, however, was contested by various nature conservation organizations and interest groups across the world. The lack of consistency at the global scale persisted until 1948, which was marked by the establishment of the International Union for the Protection of Nature (today known as the International Union for Conservation of Nature – IUCN). This body quickly became the key worldwide actor for wildlife protection and protected area governance (Phillips 2004).

At the same time, the number of national parks continued to increase at a rapid rate, having only been interrupted by the two world wars. The end of the Second World War saw a new groundswell in the designation of national parks, with more than 200 such areas created in 39 countries during the 1940s (Brovko and Fomina 2008). The spatial expansion and increased popularity of national parks underscored the need for establishing a more specialized organization tasked with addressing issues specifically related to their foundation, operation and development. As a result, the Commission on National Parks (presently known as the World Commission on Protected Areas, or WCPA) was established as part of the IUCN in 1958. The Commission compiled the 'World List of National Parks and Equivalent Reserves' for the First World Conference on National Parks, held in Seattle in 1962 (Dudley 2008) – an event which helped open vigorous discussions on issues such as the implications of human activities on wildlife, species extinction, the religious and scenic values of national parks, as well as tourism development.

It should be noted that there were significant differences between the 'pioneer' parks, on the one hand, and those created in the second half of the twentieth century, on the other. While the former were established in marginal and sparsely populated areas, the latter were often located in the proximity of developed regions, where anthropogenic influence is strongly felt. This is exemplified by the fact that numerous national parks were created in the industrialized countries of western Europe, including Britain, France, Germany, Spain and Italy. The average size of the new national parks tended to be smaller, aside from exceptions such as large national parks created in the Amazon region in Brazil (Chape et al. 2008). Recognizing the discrepancies between the areas protected as 'national parks' at the global scale, a standardized definition was proposed at the IUCN General Assembly in New Delhi in 1969. According to this definition, a 'national park' was supposed to be:

> … a relatively large area that is not materially altered by human exploitation or occupation and where the highest authority of the country has taken steps to prevent or eliminate exploitation or occupation in the whole area. (IUCN 1969 cited in McNeely 1994: 392)

The assembly made a global appeal for countries not to use the category 'national park' to name protected areas that did not meet these criteria (Phillips 2004). Yet a new set of requirements were added only a few years later, including: a minimum size of 1,000 ha; statutory legal protection; as well as a budget and staff sufficient to provide effective protection and prohibition of exploitation of natural resources. Nevertheless, the numbers of 'real' and 'quasi' national parks – at least according to the IUCN's own definition – continued to increase. Modifications to the concept included areas that were very commercialized with a focus on tourism development, as well as territories that were managed as strict nature reserves (Brovko and Fomina 2008). As a result of the former, the overall number of tourists in national parks saw a rapid rise. The effects of these trends on local ecosystems – as well as issues related to visitor management, such as raising environmental awareness – became key discussion points at the Second World Conference on National Parks, held in Yellowstone in 1972 (Dudley 2008).

The creation of national parks gradually became associated with broader social, political and environmental issues. During the 1970s and 1980s – a period marked by the invention and popularization of the 'sustainable development'[1] concept – IUCN focused on synchronizing environmental goals with social justice and economic development priorities within protected areas (Barraclough 2001). Indeed, such questions were a major discussion point at the Third World Congress on National Parks, which took place in Bali in 1982. They reflected the zeitgeist of the 1970s, which was one of the most active decades in human history with regard to the adoption and implementation of environmental policies worldwide. Thus, in 1970, President Nixon signed the first US National Environmental Policy Act; UNESCO's Man and Biosphere Programme was initiated in 1971; the World Heritage Convention took place in 1972; in 1973, EU adopted the first Environmental Action Programme; the Ramsar Wetlands Convention was enacted in 1975; the European Union sanctioned the first version of the Bird Directive in 1979.

As a whole, developments during the 1970s created a situation whereby 'national parks' became venues for balancing conservation and development needs (Chape et al. 2008). As a result, the outcomes of the Third World Congress on National Parks recommended that extant management practices should be changed, by prioritizing people-related aspects and the intensification of co-operation with indigenous and other local communities (IUCN 2010). The Congress was not only significant for the recognition of the need for 'co-management' of national parks, but also for the suggestion that all countries should increase the total size of protected areas to at least 10 per cent of their territories (McNeely and Miller 1984). All the proposals from the Congress were summarized in the well-known Bali Action Plan.

The existence of different nature conservation policies at a range of scales and via different institutional mechanisms – including conventions, declarations

---

1   The term 'sustainable development' became widely known after the World Commission on Environment and Development's (WCED) publication titled 'Our Common Future' in 1987, also known as the Bruntland report after the then prime minister of Norway, Gro Harlem Brundtland. The WCED used the term as a unifying theme in presenting its environmental and social concerns about worrisome trends toward accelerated environmental degradation and social polarization in the 1970s and 1980s. It stated that 'sustainable development is development that meets the needs of the present without compromising the ability of future generations to meet their own needs. ... Even the narrow notion of physical sustainability implies a concern for social equity between generations, a concern that must be logically extended to equity within each generation' (WCED 1987: chapter 2, paragraphs 1 and 3).

and agreements – constantly modified the meanings and functions of national parks. For example, UNESCO's Man and the Biosphere [sic] Programme (later supported by the Seville Strategy prepared in 1995) stressed that national parks cannot be only seen as isolated protected areas, but should be integrated into larger networks of biosphere reserves supporting wildlife and human wellbeing. This landscape-integrating approach was internationally discussed at the Fourth Congress on National Parks and Protected Areas that took place in Caracas in 1992. In line with the practice established at previous gatherings of this kind, the findings from the Congress were synthesized in the Caracas Action Plan in the form of recommendations and support for the biosphere reserve concept.

Thus, the sustainable development and landscape approaches initiated in the 1970s were the main driving policies in national park management during the 1980s and 1990s. The growing commitment to sustainability (especially after the Earth Summit in Rio in 1992) provided the basis for one of the most important documents regarding nature conservation: The Convention on Biological Diversity, whose Article 6 obliged the signatory countries to:

> Develop national strategies, plans or programmes for the conservation and sustainable use of biological diversity ... (Secretariat of the Convention on Biological Diversity 2001: 89)

The convention's entering into force in 1993 created the first international mechanism for monitoring the state and use of biodiversity at the global level. It should be added that the management of national parks and their biodiversity became part of the national monitoring strategies as well.

Another important step was made in 1994; This was the year when IUCN created the system for categorizing protected areas that is currently in force (Dudley 2008). The new classification framework was developed in accordance with the objects of protection, rather than 'who owns, controls or manages a particular protected area' (ibid. 25). Based on the primary management objective and the degree of human intervention, the system incorporated six categories of areas mainly protected for:

1. Strict protection (involving '1a – Strict nature reserve' and '1b – Wilderness area');
2. Ecosystem conservation and protection (i.e. national parks);
3. Conservation of natural features (natural monuments);
4. Conservation through active management (represented by habitat or species management areas);
5. Landscape/seascape conservation and recreation;
6. VI Sustainable use of natural resources (such as managed resource protected areas).

Overall, this categorization foresaw the role of national parks as one aiming:

To protect natural biodiversity along with its underlying ecological structure and supporting environmental processes, and to promote education and recreation. (Dudley 2008: 16)

It should be noted that areas with conservation aims and governance practices that were significantly different than those prescribed by the IUCN system still continued to be named as 'national parks' across the world, subsequently becoming linked to other protection categories in the framework. In part, this is due to the fact that many countries defined an array of criteria – mainly aimed at reflecting what they considered the 'public interest' – in order to determine the representativeness or the vulnerability of landscapes intended for protection. However, valuation processes often involved positivistic technical and data-driven approaches, such as determining the number, importance, origin and abundance of species or geomorphological features. Combined with local conflicts of interest and the lack of data, such thinking often led to biased decisions.

According to Watson et al. (2011), the most effective biodiversity protection practices over the last few decades have been based on a few key principles, including: the identification and protection of representative habitats; the delimitation of objects of protection; the protection of hotspot habitats for endangered species; and the establishment of efficient governance against threats to protected sites. But experts and practitioners alike have increasingly critiqued the prioritization of nature protection over the wellbeing of local people. As a result, the late 1990s and early 2000s saw the emergence of a range of more inclusive co-governance[2] models (Ostrom et al. 1994).

Governance-related issues received particular attention at the Fifth World Parks Congress held in Durban in 2003. Over 120 representatives from indigenous, mobile and local communities participated in this event. Some of the most important decisions endorsed at the Congress included the acceptance of pluralist co-governance practices as models for expanding the global protected area network, and increasing its legitimacy and efficiency (Chape et al. 2008). Even though co-governance models were also recognized by the Convention on Biological Diversity, they have provided insufficient cover against the new challenges faced by national parks. The last three decades have been marked by a mounting body of evidence in favour of anthropogenic climate change (Treasury 2007, Parry 2007), with significant implications for nature and humans (Peters 1985, Staple and Wall 1996, Bartlein et al. 1997, Scott et al. 2002a, Heller

---

2　Co-governance (sometimes referred to as participative, joint or multi-party governance as well as co-management) is commonly understood as a governance model that includes two or more subjects (i.e. governmental authorities, non-governmental organizations, private parties as well as community and indigenous groups) which negotiate, define and share governing functions and responsibilities in order to achieve nature conservation and social equity in a particular protected area (Mappatoba and Birner 2004).

and Zavaleta 2009). This has necessitated further modifications to governance models, in order to enable national parks to cope with the new uncertainties.

In light of such circumstances, it remains unclear how national parks can become more adaptive and resilient to future disturbance, stress and shocks. Watson et al. (2011) have argued that the 'static' nature of the principles that have commanded hitherto biodiversity protection practices is being problematized by climate change. In new planning paradigms, the words 'adaptive'[3] and 'resilient'[4] have become key driving factors in defining the coping policies of national parks. However, managing organizations are still uncertain about the specific measures that need to be enacted in order to prepare national parks for climate change. This is linked to the open question as to whether adaptation should be a matter of responding to climate change as it manifests over time, or whether anticipatory or precautionary measures should be devised in order to make parks resilient to climate change effects (Baron et al. 2009, Lemieux et al. 2011).

In response to such dilemmas, academic scholars and conservation practitioners (Scott et al. 2002b, Lemieux and Scott 2005, Cross et al. 2012, Stein et al. 2013) have proposed a plethora of adaptive governance and conservation activities: improving interdisciplinary cooperation among the interested parties; raising awareness about the risks among local communities; increasing the size and connectivity of strictly protected areas (under category 1 above) and vulnerable habitats by creating safe refugia; reducing non-climate threats (such as habitat fragmentation, pollution and poaching) as well developing bioclimatic prediction models. However, in their review of such strategic plans and recommendations, Heller and Zavaleta (2009) have found that most of them neglect social science issues, and are almost entirely focused on ecological data. They also argue that the formulation of inclusive and holistic approaches over the past 20 years has often failed to pay sufficient attention to the significance of conservation and restoration in human-dominated landscapes (also see Peters 1985).

The uncertainties brought about by climate change have combined with efforts to make wildlife and habitats more resilient to yield a strategy known as 'rewilding' – sometimes termed 'ecological restoration' – which has been gaining increasing prominence over the past two decades. Its constituent 'cores, corridors, and carnivores' paradigm is based on two principal pillars: i) expenditure and connectivity of core-protected areas with corridors that enable humans and wildlife to co-exist; and ii) protection and reintroduction of currently extinct species to areas from which they were extinct in recent history. Michael Soulé – the founder of conservation biology – and his colleague Reed Noss have been among the

---

3   The basic premise of *adaptive management* is 'learning by doing' and taking actions relying on what was learnt (Walters and Holling 1990, Williams 2011).

4   Resilience management can be defined as a structured approach, which ensures 'functionality of the system when it is perturbed, or maintaining the elements needed to renew or reorganize if a large perturbation radically alters structure and function' (Walker et al. 2002, available at: http://www.consecol.org/vol6/iss1/art14).

initiators of this idea, which has been touted as a more inclusive and integrated concept than the species reintroduction practices that have been applied since the mid-nineteenth century.

Operating under the simultaneous guise of a conservation method and a grassroots movement, rewilding has become an expanding global phenomenon. Its materializations have been visible at an extremely wide range of scales: from small habitat restoration practices such as the return of fens to the UK, to the reintroduction of apex carnivores such as the wolf, bear and lynx to EU Natura 2000 areas, as well as the return of wolves to Yellowstone and the repopulation of Africa with south China tigers. The procedure has also involved the creation of large wilderness areas via the use of non-intervention management practices in transboundary national parks. As a whole, rewilding has often been driven by the belief that the human-led extinction of the Pleistocene megafauna led to an imbalance in global ecosystems, making them more vulnerable to the possible implications of climate change.

Rewilding programmes have been surrounded by multiple controversies as a result of their varying success rates and implications for local communities. Selected conservation activists and natural scientists have invested significant amounts of energy in attempts to justify such strategies in economic and social terms. It has been argued, for example, that the reintroduction of wolves to Yellowstone National Park both improved the natural ecological balance and increased the total annual economic and social benefits by 6 to 9 million US dollars (Duffield et al. 2008, Sandom et al. 2012). The Russian scientist Sergey Zimov suggested that the ecologically 'damaged' Siberian landscape can be 'repaired' and made more resilient to climate change by reintroducing the once abundant wild horse.

The Purpose and Character of National Parks: A Global Survey

Despite the broader trends identified above – pertaining to the use of national parks for functions beyond nature protection and conservation – it is also clear that the borders, management and meanings of such areas have been shaped by the specificities of place and space. Therefore, any discussions of the involvement of local populations in the governance of national parks must take into account the wide geographical variations in the role and character of these territories.

The US and Canada: Birthplace of the National Park

National parks have been termed 'America's best idea', having been established in order to provide 'public pleasure and benefit' in terrestrial or water areas of 'supreme scenic splendour or other unique qualities', without threatening the attributes that can be enjoyed by future generations (Stegner 1987). Indeed, The US Congress created Yellowstone – the world's first national park – in order to serve as 'a pleasuring-ground for the benefit and enjoyment of the people'. It is important to note, therefore, that such areas were principally created to provide scenic and recreational amenities, rather than supporting and conserving biological species and ecological processes (Westman 1990, Lorah and Southwick 2003, Merenlender et al. 2004). National parks were instituted on what was presented as 'empty' federal land in the US, although many groups of Native Americans used these areas as their homes, hunting grounds, and transportation routes for thousands of years.

However, the main perceived challenge in the establishment of US national parks was not the presence of indigenous local groups – as their rights to land were not really taken into consideration at the time – but the pressure from the business community. As argued by Sellars (1997), the growing commercial benefits brought about by national parks led to increased demands by developers and industrialists. This meant that economic 'worth' became an important factor in the decisions about such areas made by US Congress (ibid.). The notion of worth has been explained as:

> ... the persuasiveness of economic arguments in determining precisely which scenery the nation felt it could afford to protect in perpetuity. As I originally explained, the term "worthless" grew out of the congressional debates. The word consistently referred only to the absence of natural resources of known commercial value, not to scenery, watersheds, or wildlife with obvious inspirational or biological – if not direct monetary – worth. (Runte 1997)

Nevertheless, the activities of a strong lobby of prominent nature conservation advocates (such as Frederick Law Olmsted, John Muir, and George Bird Grinnell) has led to the creation of 58 national parks since 1972; the latest example is the Great Sand Dunes National Park, which was established in 2004. All the parks are included in the National Park System, which is managed by the National Park Service (NPS), founded by the Congress in 1916 as an agency of the Department of the Interior. According to the US Organic Act from 1916, the NPS's role is to promote and regulate sustainable use of US national parks for 'enjoyment' and to conserve their 'natural attributes'. The dual mandate of NPS to encourage both tourism and conservation, however, has been an inherent source of conflict. NPS has also been criticized for prioritizing enjoyment and tourism related goals (its declared goal 'A' is that parks act as 'places for people to enjoy') over conservation objectives (the rather imprecise goal 'B' states that the NPS will 'improve ... park resources and assets') in its centennial strategy (Antolini 2008: 883).

Canada was among the first countries to copy the US 'national park' concept in 1885 by protecting the land around the Banff hot springs on Sulphur Mountain from 'sale, settlement or squatting' while providing a public space for 'benefit, education and enjoyment' (Legislative Services Branch 2013). The national park system consists of 'representative' national landscapes. It was also the first country to create a national agency – the Canada Parks Agency (CPA) – responsible for national park management in 1911. Pristine natural areas within national parks were especially protected as wilderness areas in the 1930s.

Nevertheless, the US and Canada had relatively similar national park trajectories at the beginning of their park histories. As a result of the common conceptualization of national parks, the governments of both countries established the first international peace park in 1932, by uniting two adjacent national parks: the Waterton Lakes National Park (Canada) and Glacier National Park (US). The park was created in the name of 'peace and goodwill' between the two nations.

In Canada and the US alike, one of the prime arguments for creating national parks revolved around the provision of unique opportunities for acquainting the public with nature. The rising number of visitors to the parks was perceived as a direct reflection of public and political support for such areas (Shultis and More 2011). Yet it should be noted that neither US nor Canadian parks were popular tourist destinations in the period immediately after their establishment. It was the development and expansion of railway networks in the remote regions of North America that provided the main impetus for visitors to escape polluted urban areas and look for scenic sanctuaries in such areas. The number of tourists subsequently became proportional to the perceived popularity and accessibility of national parks.

Given their public role and character, national park authorities were not immune from wider political developments. As a result of the rise of global neoliberalism in the 1970s (Harvey 2007), US national parks began to be considered as service providers, which had to be 'marketized'. Their funding from public sources was subsequently transferred onto a market-based operating system (Shultis and More 2011) making them dependent on user fees, marketing, outsourcing, and public-private partnerships (Crompton and Lamb 1986, Lehmann 1995, Crompton 1998 cited in Shultis and More 2011). Tourism became the single most important financial contributor to the national parks' budgets with customer satisfaction acting as an 'indicator' for success. In order to satisfy the needs of visitors, many parks significantly expanded the services and commercial opportunities provided in this regard (Lowry 1994 in Shultis and Moore 2011).

Not only were such trends significantly less pronounced in Canada, but national park management between the two nations also diverged with regard to the level of centralization of local and regional government. While Canada undertook significant efforts to devolve central state power to sub-national scales of governance, the bulk of US regulatory and governing authority remained concentrated in Washington (Lowry 1994). This high degree of centralization in the latter increased the level of political control in the management of national parks, mainly due to the frequent appointment of officials focused on short-term gains that benefit the particular constituencies. At the same time, the power shift away from the central state in the former led to the emergence of a preservation-minded professional bureaucracy, with ecological integrity becoming the cornerstone for the existence of the national parks (Parks Canada Agency 2000).

Although parks were open for visitors and admirers, both countries managed these territories in an exclusive and top-down manner until well into the 1990s (Murray and King 2012), when the existence of 'new' participative approaches was recognized, particularly in relation to the rights of, and need for, the involvement of local communities. The new model led to the signing of numerous co-operation treaties with local people; one of the most notable examples is the British Columbia Treaty from 1993.

The border areas of the national parks also became attractive for residential and business development. Hansen et al.'s (2011) report that the counties surrounding the large national parks in the western parts of the US – particularly Yellowstone and Grand Teton – belonged to the top national tenth percentile in population growth from 1970 to 1997. Moreover, the number of housing units within 50 km of national park areas has increased by an average of 57 per cent for every decade from 1940 to 2000; this is well above the national average of 21 per cent (Radeloff et al. 2010).

A further distinguishing feature of US national parks can be found in the significant differences between the size and character of such territories in the western and eastern parts of the country. While the former tend to be larger, less easily accessible and located in more remote and mountainous areas (notable examples include Glacier, Yellowstone, Grand Teton, Rocky Mountain and Yosemite) the latter are smaller, surrounded by more private land and characterized by proportionally higher development rates, partly as a result of being subjected to a longer history of human settlement in their surroundings (this is best illustrated by the situation in the Great Smoky Mountains, Big South Fork, Shenandoah, New River Gorge, Delaware Water Gap, see Parks and Harcourt 2002, Wade and Theobald 2010). Still, even the remote western national parks feature higher-than--average increases in population and housing (Davis and Hansen 2011). Many of them have also struggled to maintain or reintroduce top predators as a result of human-wildlife conflicts with adjacent private properties, especially as a result of damage inflicted by wildlife and poachers. Such areas have also been affected by large development projects, such as, for example, wind energy farms in the surroundings of Mojave and Glacier, and logging on public land on the borders of Mount Rainier and Crater Lake.

It should also be noted that the US National Park System has a complex institutional structure. It contains additional categories of national parks and their equivalents: historic sites, recreation areas, wild rivers, and scenic trails. The Wilderness Act of 1964 resulted in the creation of a National Wilderness Preservation System, which incorporates federally owned lands designated by Congress as 'wilderness areas'. These have been defined as 'areas where the earth and its community of life are untrammeled by man, where man himself is a visitor who does not remain' (US Congress 1964: 1). Many of such areas have been created within the boundaries of extant national parks, transforming them into Wildland protected parks; notable examples include Yellowstone and Mount Rainier.

Latin America: Rich Biodiversity, Numerous Stakeholders, Large Parks

Latin America's first national parks were created immediately after the establishment of their counterparts in the US and Canada. Instituted in Mexico as early as 1882, El Chico National Park is considered to be the first such protected area in this region; it was followed by examples from across Central and South America. Until the early twentieth century, these territories were set up in line with the 'pristine – scenic' model: basic ecological concepts such as 'biodiversity' or 'ecosystem integrity' were not included in the rationale for their conservation. As a result, the first national parks were created in areas considered aesthetic, pristine and valuable for the development of tourism (Pauchard and Villarroel 2002). However, the need to preserve 'pristine nature' also meant that many local communities were destroyed, with their inhabitants relocated outside the parks' boundaries. Some of the social and political problems resulting from such evictions are still felt today.

In countries such as Argentina, the creation of the first national parks often reflected the interests and desires of prominent naturalists. The country got its first national park in 1903, when Francisco Moreno – a well-known Argentinian explorer – donated his land to the government (Aagesen 2000). In Chile, forest regions were seen as possessing greater aesthetic value, which meant that they were prioritized over potentially more ecologically valuable areas. Designating national parks in woodland territories was also seen as a method of stopping the rapid deforestation caused by agricultural activities and land settlement (see Figure 3.1). Indeed, as argued by Pauchard and Villarroel (2002) for Chile:

> … the northern desert region, with distinct ecological and scenic values, was apparently not considered a priority for protection, probably because it lacked forests or other lush vegetation. (321)

The boundaries of Chilean parks, however, have often been established and contested in relation to multiple social, economic and political trade-offs. High land values and dense settlement patterns have been identified as the main obstacles for the protection of unique and diverse ecosystems in central Chile.

National parks in different guises continued to be created throughout the twentieth century. Most notably, this trend included the foundation of the Kaiteur National Park in Guyana (1929), the Galapagos National Park in Ecuador (1936), and the Sajama National Park in Bolivia (1939). Argentina's Law on Protected Areas, adopted in 1934, conceived national parks as a means of solidifying national sovereignty in frontier regions. Four such territories were created along the southern border with Chile in 1937, in order to 'protect' the country from foreign incursions. Most extant residents were classified as Chilean 'intruders', and as such were expelled by force (Aagesen 2000). The rights to land and resources of those who managed to stay were significantly curtailed, leading to the rise of unsustainable land-use practices (Amend and Amend 1995).

The 1940s and 1950s saw significant changes in the management policies of many Latin American national parks (Hopkins 1995, Hillstrom and Hillstrom 2004). Although property and capital rights were still very limited, local people were not expelled any longer. Many residents, however, were charged for permits to use the parks' resources in a controlled manner. Nevertheless, the number of national parks continued to grow at an exponential rate, especially after the 1960s.

Most governments defined national parks as strictly protected areas aimed at conserving ecological integrity and diversity, cultural features, and scenery. Such spaces, therefore, were primarily reserved for tourism, education and scientific research. Direct and indirect income from visitors – particularly those from overseas – thus became a key source of revenue across the region. In many cases, tourism started to shape the economies of settlements and regions adjacent to the parks. For example, tourist services replaced the dominant mining industry in the Chilean town of Puerto Natales as a result of the rapid rise of visitor numbers in nearby Torres del Paine National Park. According to Villarroel (1996) one in nine families in Puerto Natales sources its income from tourism-related activities, resulting in gaps in employment due to the seasonal character of the work. An even greater problem has been posed by the insufficient involvement of local people in the management and development of tourism; the locals' limited access to education and capital has affected visitor numbers (Clapp 1998). Such examples abound from across the continent, despite the rising importance – especially in terms of economic income – of nature- and ecotourism-orientated activities in national parks.

Overall, the lack of adequate access to land and ownership rights has affected the management of national parks throughout southern and central America. Most protected areas in this region were created on what was considered 'national/ public' land allegedly owned by the state. However, there is evidence to suggest that many of the parks extend over territories occupied by indigenous people for generations; this is especially true in the Amazonian region in Brazil. There are a few examples, however, where lawmakers have addressed this issue; most notable is the recognition of the Maya communities' rights to land in the Sarstoon Temash and Rio Blanco national parks in Belize's Toledo District. After a long political struggle, the land titles were returned to the two Maya communities in 2007 (Ch'oc 2012).

In recent years, the expansion of climate change and CO_2 emissions trading systems has led to a subtle transformation in the role of protected areas in the region, mainly thanks to the presence of abundant forest resources. CO_2 sequestration in particular has turned the national parks located in South America's rainforests – particularly in the Amazon basin – into key market-players in the international emissions trading system. Numerous high-level political negotiations and initiatives have grown out of the perception that providing carbon-market-based incentives (such as REDD and REDD+) can help slow down deforestation processes and promote economic growth. For example, the Ecuadorian President Rafael Correa stated in 2007 that his country would limit the profits from one of its most abundant oil reserves in the Amazonian Yasuni National Park in exchange for 'reasonable' compensation from the international community for keeping the oil underground and protecting the forest. The compensation amount was estimated to equal at least half of what Ecuador would have received from large oil companies – Enlap, Sinopec, and Petrobras – for the oil (Martin 2011).

Local people have often been sceptically inclined towards 'green capitalism' programmes. In October 2011, the participants at the workshop '*Environmental Services, REDD and BNDES Green Funds: The Amazon's Salvation or a Green Capitalism Trap?*' held in Acre (World Rainforest Movement 2011), wrote a letter that criticized REDD and the commodification of nature:

> Although it is presented as a solution for global warming and climate change, the REDD proposal allows the powerful capitalist countries to maintain their current levels of production, consumption and, therefore, pollution. They will continue to consume energy generated by sources that produce more and more carbon emissions.

The letter (ibid.) also stated that programmes such as REDD promote and combine:

> … two forms of re-territorialization in the Amazon region. On one hand, it is evicting peoples and communities from their territories (as in the case of mega projects like hydroelectric dams), stripping them of their means of survival. On the other hand, it is stripping those who remain on their territories of their relative autonomy, as in the case of environmental conservation areas. These populations may be allowed to remain on their land, but they are no longer able to use it in accordance with their ways of life.

Despite the multiple tensions and challenges, there is little doubt that national parks remain one of the key components of environmental and economic governance frameworks in Latin America. This is evidenced by the fact that approximately 20 per cent of South American nature is protected (Clark and Aide 2011). Some countries from the region have placed around one third of their territories under protection (i.e. Belize 38 per cent, Panama 29 per cent, Guatemala 28 per cent, Venezuela 27 per cent, Costa Rica 26 per cent). Over 30 per cent of these protected areas are not categorized (Dudley 2008).

Africa: Struggling with the Colonial Past

The development of Africa's national park system has been deeply shaped by the divergence between the experience and practices of its colonial history, on the one hand, and the post-colonial period that followed, on the other. During colonial rule – and in line with early twentieth century Western perspectives on conservation – national parks were created principally in order to preserve nature in all its splendor 'for eternity' (Dunn 2009: 436). As a result, the main objective behind the establishment of the first national parks on this continent – Albert National Park in 1925, and Odzala National Park in 1935 – was the conservation of large areas of dramatic wildlife.

Early national park governance efforts were characterized by a disparaging attitude towards traditional conservation practices, such as the unofficial system of sacred trees and groves. Moreover, the borders of national parks were determined without taking into consideration local political and cultural sensitivities (Chape et al. 2008). National parks were often claimed as Crown property, leading to the eviction of thousands of local inhabitants (Hulme and Murphree 2001, Brockington et al. 2006, Büscher and Dressler 2012). Most of the displaced people were settled in areas along the national park boundaries, which meant that they could continue their traditional hunting and agricultural activities. However, since hunting was banned, previously common everyday practices started to be considered criminal, and local people were seen as poachers (see Brockington and Igoe 2006). This was aided by the fact that national parks were managed in a rigid and centralized manner, most often by forest departments established by colonial governments.

The majority of African states gained independence in the period between the 1950s and 1970s. Even though the post-colonial era was eagerly anticipated as an opportunity for Africa to determine its own development and governance paths, colonial nature conservation concepts continued to persist. Not only was there little substantive change in the national parks' management practices, but they also continued to be heavily dependent on international conservation advice and funding. The involvement of foreign practitioners facilitated the hierarchical and top-down management structures established by post-colonial governments (Adams 1992, King 2010, Nelson 2012). As argued by King (2010):

> The transition to the postcolonial era has resulted in expanded, rather than reduced, control of local landscapes and peoples by external actors. (5)

The colonial practices of displacement and expropriation of private land thus continued in the post-colonial era. Local people often paid the biggest price for this. There is widespread evidence to suggest that communities living in the immediate proximity of national parks have suffered from record poverty and mortality rates (de Sherbinin 2008), mainly as a result of limited opportunities for economic and social development. The wellbeing of African national park populations has also been affected by frequent civil unrest: protected areas in Congo, Liberia and Sierra Leone have faced unprecedented settlement pressure from refugees originating in conflict areas. This has prompted some governments to consider national parks – especially those along national borders – as places for affirming the sovereignty of the central state.

Thus, the main distinguishing features of post-colonial national park governance in Africa do not necessarily stem from the management practices in them, but from the consequences of wider social, economic and political developments. This is illustrated by Dunn's (2009) claim that African national parks represent 'contested state spaces'; their multiple roles – wildlife protection, homes, tourist destinations, refugee camps – have turned such areas into places:

> … where officially sanctioned state-making practices are successfully challenged, resisted and replaced by alternatives. (423)

The 1990s brought about a significant shift in the relationship between African national parks and local communities. International pressure in favour of communities' rights to land and resources led to the restitution of territories seized from evicted populations. One of the first such developments took place in 1999, when 20,000 ha from the Kruger National Park in South Africa were returned to the Makuleke people. Despite predictions that the park would be destroyed, the area is currently jointly managed by the Makuleke people and the responsible state agency – South African National Parks (SANParks) (Reid 2001, Robins and Waal 2008, Cundill et al. 2013). Community conservation management practices have also gained increasing popularity in recent years: one of the best known examples can be found in the Communal Areas Management Programme for Indigenous Resources (CAMPFIRE) undertaken in Zimbabwe and South Africa (Child and Barnes 2010, Gandiwa et al. 2013). However, post-implementation impressions for many such projects have curbed previous enthusiasm, as it has often transpired that they *de facto* aid neoliberal top-down policies, with unclear ecological, economic and social benefits (King 2010).

The realization that the long-term viability of African wildlife hinges on overcoming the isolation and fragmentation of protected areas (see Figure 3.2) has led to the creation of a continuous network of transfrontier conservation areas (TFCAs, or 'peace parks'). The first such area was Kgalagadi Transfrontier Park, established in 2000 by merging the Kalahari Gemsbok National Park in South Africa and the Gemsbok National Park in Botswana; it encompasses an area of approximately 3.6 million ha. The increasing popularity of the peace park model has led to the creation of several such territories across the continent; one of the most notable examples is provided by the Kavango-Zambezi Transfrontier Conservation Area, formed by the governments of Namibia, Angola, Zambia, Botswana, and Zimbabwe in 2006.

The creation of peace parks has often been driven by the argument that supra- and transnational forms of environmental governance can be more effective than national management mechanisms (Wolmer 2003). Transfrontier parks have also been seen as a means for maintaining transboundary cultural and ecological integrity in Africa; it is believed that they can redress historical injustices and imbalances created by the division of previously compact African ethnic communities and ecosystems by national boundaries (for example, the border between Mozambique and Zimbabwe separated the Barwe, Ndau, Manyika and Shangaan communities).

Even though peace parks have been perceived as emblematic of the post-colonial 'African Renaissance', they have also received a significant amount of criticism. It has been observed that TCFAs have often failed to avoid conflicts and controversies in Africa, as a result of the inability to address clashing national interests as well as 'insufficient community consultation and sensitive border issues such as the illegal flows of goods and migrants' (van Amerom and Büscher 2005: 1). Duffy's (2006) concerns are especially pertinent in this regard:

> ... the practice of TFCAs falls far short of their promise ... The assumption that management of transboundary environmental problems through forms of global governance which utilise a technical or scientific rationale, and which adhere to neoliberal political and economic principles, is deeply problematic. (109)

Asia – Parks as Reserves and Amusement Playgrounds

The pre-modern idea of protected areas – including 'parks' – has existed in several Asian societies for thousands of years. In China, park-garden types of protected areas were created more than 2000 years ago, during the rule of the Qin Dynasty; they mainly served aesthetic purposes, and were located either on private land or in the vicinity of temples and burial sites (Edmonds 2002). The *henna* areas in the Middle East – large land parcels traditionally protected mainly to prevent overgrazing (Chape et al. 2008) – can also be classified in this category.

Even though Asia has a diverse history of modern park governance, the factors that have shaped their establishment are geographically distinctive, and common to many states across the continent. Only a few Asian countries participated in the international euphoria for creating national parks at the beginning of the twentieth century. They included Japan, where, most notably, the Japanese movement of national park supporters propagated the creation of parks as a means of commemorating the semicentennial jubilee of Meiji Emperor's reign. Although the project was abandoned due to the Emperor's death, its underlying idea was taken forward and spread by the general public, thanks to the belief that establishing a national park can bring prestige to its host region (Hiwasaki 2005).

A much broader wave of establishing national parks across south and east Asia commenced in the 1930s. With the exception of Brunei and Thailand, south Asia's parks from this period were predominantly products of colonial nature conservation policies, as most of them were established by transforming royal hunting reserves. One of the best examples is provided by the Jim Corbett National Park – created in India in 1936 – which pioneered these principles at a grand scale. The adoption of Japan's first Law on National Parks in 1931 resulted in the establishment of 12 parks by 1934, mainly with the aim of promoting scenic landscapes and attracting tourists. It was believed that the national parks would also enhance the country's prestige on the international scene, and help build patriotic feelings (Hiwasaki 2005). World War II further accelerated the process of creating new national parks, as it was believed that they would assist the development of tourism and provide badly needed financial resources for the reconstruction of the country. Thus, the first parks mainly served recreational purposes, and areas such as mountains and forests were not included within the nature protection system.

The first national parks in the rest of the continent were created between 1960 and 1980: they include, most notably, Khao Yai National Park (Thailand, 1962), Chirisan National Park (Korea, 1967) and Royal Chitwan National Park (Nepal, 1973). It is worth noting that Vietnam's first national park – Cuc Phuong – was established in 1962, when the country was in the middle of a war. Countries such as the Maldives designated their first national parks as late as the 1990s.

Asia's largest country in population terms – China – currently has a highly specific and complex nature protection system, including a range of national park categories. The fact that its national parks have been designated by different ministries means that a unified national park system is lacking (Liu et al. 2010). Closest to the IUCN's second category are 'national forest parks'; the first such area – Zhangjiajie – was established in 1982 (Wang et al. 2012). Yet these territories are seen as 'paper parks' by IUCN, due to being established mainly for development and tourism purposes, rather than nature conservation (Xu and Melick 2007). Thus, it has been reported that the government of Sichuan Province leased one national park to a private company to operate with it for over 50 years. In response to such criticisms, the Pudacuo National Park – designated by the Yunnan provincial government – asserted to be 'the first "real" national park in China' in 2007. Its claim has been challenged by Tangwanghe National Park, which was established by the Chinese Ministry of Environmental Protection in 2008, by applying American nature conservation principles.

It is also worth noting that land ownership patterns in national parks vary significantly across Asia. In most countries (including those in the Middle East, as well as Mongolia, North Korea and China) these areas are owned by the state. Conversely, Japan allows private national parks to be created based on personal initiative. After being designated by the Minister of Environment, however, the parks must be managed in accordance with the regulations on land use and development. The areas may continue to be inhabited, and people might continue agricultural and forestry activities throughout their territory, except in core areas (Hiwasaki 2005).

While some parks in Asia have accepted local people and communities as an integral part of the landscape, others have managed natural resources in an exclusionary manner. Even though recently established national parks in Asia have been managed in a more inclusive way, indigenous and traditional people still face threats of eviction, and difficulties in claiming rights to resources and land. This is particularly true in south east Asia, as exemplified by the plight of Akha hill tribes in northern Thailand (Gray and McCabe 2010).

Tourism – in its multiple guises – has been propagated as a win-win economic activity that may be beneficial for nature conservation by replacing agricultural and forestry activities. As a result tourism has become the main provider of finance for millions of households throughout Asian national parks. For example, it was estimated that 60 to 80 per cent of the local population has gained some form of income from tourism related activities in Sagarmatha (Mount Everest) National Park in Nepal (Nepal, 2002). At the same time, tourism has also been identified as a trigger for industrial development in national park areas – especially in countries such as China, where economic growth has been rapid. However, this activity has also been singled out as the main threat for wildlife destruction, prompting some Asian countries to adopt new legislation and policies to regulate it. For example the Ministry of Forestry in India – the institution responsible for the national parks – decided to close most of the tiger reserves in the country in July 2012, in response to claims that the presence of tourists traumatizes the tigers. The decision has endangered the livelihoods of thousands of families dependent on tiger-related tourism.

Australia and New Zealand: Parks Equal National Identity and Freedom

The developments that led to the early establishment of national parks in the US and Canada were also felt on the other side of the Pacific: in New Zealand and Australia. The Royal National Park – the first such territory in Australia and second worldwide – was created in 1879. The area's main purpose was to serve as a recreational backyard for the inhabitants of Sydney, rather than protecting the natural landscape. Nevertheless, the foundation of this park triggered the establishment of numerous similar institutions across the country. Further aiding the process was the need for managing forest products – mainly timber – and developing tourism. Only in the 1970s did Australia set up separate park authorities, thus removing parks from the management of forest services (Herath 2002). This was accompanied by a shift of the protection regime towards wildlife and ecosystem conservation (Buckley 2003, also see Figure 3.3).

According to the General National Parks and Wildlife Act from 1975, Australia has eight systems of national parks and reserves, only one of which has national status (the remainder are regional). The national parks located in the six Australian states are created and governed in accordance with nature protection policies developed individually by each state. While all national parks across the country are situated on public land and are primarily funded from public state budgets, the parks located in external Australian territories are governed and funded by the national government. Even though the parks are managed as public places with mandated access, they are nevertheless obliged to recover costs via a system of entry fees. This has created a paradox between their public nature and the requirement for equitable public access on the one hand, and the pressure for implementing market-based principles, on the other (Hughes and Carlsen 2011).

Australia and New Zealand also possess a long tradition of co-operative management practices, as evidenced, for example, by the co-management agreement established with the traditional owners of Kakadu National Park in 1978. The agreement allowed traditional owners to lease their land to the federal government in return for an annual fee, in addition to a proportion of revenue earned from park user fees. A Board of Management with a majority of local owner representatives was tasked with the preparation, implementation and monitoring of a management plan for the reserve. It should be added that three of the six federal national parks – Kakadu, Uluru-Kata Tjuta, and Booderee – are also co-managed by the Department of Environment and Heritage in conjunction with traditional Aboriginal owners.

In New Zealand, the Tongariro National Park – the country's first – was established by the Ngāti Tūwharetoa people, who endowed the land surrounding the peaks of the sacred Tongariro volcanoes for protection purposes to the state. Although more parks were created in the following years, the New Zealand national park system was consolidated much later, with the adoption of the National Park Act in 1952. Since 1987, national parks have been managed by the Department of Conservation alongside all other natural and historical conservation areas. Entry and access to the parks are generally free for the public, but in some cases may be specifically regulated in accordance with nature conservation activities. Until 2007, the country designated 14 national parks – mostly in mountainous and forested areas – covering approximately 11.5 per cent of its territory.

The wild areas encompassed by national parks are seen as possessing great social, ecological, cultural and economic value to the New Zealanders. 'Wild' landscapes have also become one of the main attributes defining the identity of this nation (Bell and Lyall 2002), having been used to distinguish New Zealand from England and provide a basis for the construction of a distinctive sense of national belonging following the country's independence declaration in 1947. This situation has transpired despite the fact that more than 85 per cent of the country's population lives in urban areas; a figure that is almost double the level registered in the late 1800s.

The socio-economic benefits generated by New Zealand's national parks have been investigated in several studies (Clough and Meister 1989, Booth and Simmons 2000, McCleave et al. 2006). Research of the wider influence of the Te Wāhipounamu World Heritage Area – located in the southwest parts of the country, and including four national parks – has found that tourism provides the largest share of economic gains for local communities (Molloy et al. 2000, Mize 2006). The increased number of visitors, it is argued, has led to improved local facilities, while boosting property values. However, local communities have expressed concerns about the financial losses brought about by skills shortages – mainly due to the inability to prepare sound concession and lease applications – as well as the expansion of second-home construction, and the loss of traditional lifestyles (Molloy et al. 2000). At the national scale, this has been supplemented with concerns about the possible revision of existing legislation regarding the mining of mineral deposits in protected areas.

Europe – Parks as Fenced Gardens and Remote Wilderness

The trajectories followed, and current challenges faced, by Europe's national parks are very much a product of the specific geography of the continent, characterized by dense population settlement patterns and a long history of industrialization in 'core' areas, surrounded by less intensively developed 'peripheral' regions. As a result of this situation, western and northern European states found it difficult to keep up with the trend of creating national parks in the nineteenth century, as areas that would meet the 'wilderness' criteria prescribed by its underlying paradigm were few and far between. Nevertheless, Sweden, Switzerland, Spain and Italy all managed to establish their first national parks as early as 1909, 1914, 1918 and 1922, respectively. Given the global nature of the impetus to create such protected areas, they were instituted largely thanks to a top down effort. For example, Italy's first national park – Gran Paradiso – was created on land donated by King Vittorio Emanuele III to the state, mostly for conserving the local ibex population.

The need for affirming national identity also played an important role in the creation of national parks. This is exemplified by the case of Sweden – a country that found itself in a difficult position at the beginning of the twentieth century due to, *inter alia*: i) the increased mechanization of the society due to the industrial revolution, resulting in a shift of the demand for labour from rural to urban areas; and, ii) the consequent migration of numerous impoverished rural families, which left the countryside for a better life in cities, or the US. In 1905, Sweden ended its union with Norway, creating a need for symbols that would unify the new state. The protection of 'nature' – whose 'wilderness' harked back to the original 'fatherland' (Löfgren 1987 cited in Mels 2002) – was seen as a potentially crucial integrating step in this regard. As pointed out by Mels (2002), although adjectives such as 'nature', 'state' and 'protected' were all taken into consideration as designators of the newly-founded parks, none of them was sufficient to encapsulate feelings of patriotism as much as the word 'national'. When coupled with the notion of a 'park', this concept suggested that national interests were enveloping, solidifying and coalescing around a precious piece of land. The establishment of national parks thus became a prime mission for political elites, resulting in the creation of nine such territories in 1909. Their mission and management frameworks were defined in very broad terms – from protecting the genuine nineteenth-century agricultural landscape in Ängsö to conserving remote 'empty' mountainous areas in Abisco. In many cases, traditional Sami people lived in these regions, but were represented as 'part of nature': the recognition of their interests meant that they were able to maintain the right to hunt and fish.

The 1930s saw the creation of Central Europe's first national parks. The trend was started by Poland with the creation of the Pieniny and Białowieża national parks in 1932, and followed by Bulgaria and Romania in 1934 and 1935, respectively. In the decades after the Second World War, more European countries started creating national parks. In eastern and southern Europe, the ex-Yugoslavian republics of Macedonia and Croatia gazetted their first parks in the late 1940s. Nationalist discourses and folk narratives also played a key role in the establishment of such areas – this topic is treated in further detail in Chapter 4.

The UK's Labour Party defined the principles of that country's national parks in the White Paper on National Parks, which was part of the post-war reconstruction plan adopted in 1945. This document aided the establishment of the UK's first national park – Peak District – in 1951, which was soon followed by numerous similar territories across England and Wales. However, Scotland instituted its national park framework only in 2000, resulting in the creation of the Loch Lomond and The Trossachs National Park in 2002, and Cairngorms National Park in 2003. It should be pointed out that the governance of national parks in the UK is deeply affected by land-use interests, as a result of the fact that most of the land is owned by private landlords. As a result, national parks represent complex social-cultural landscapes that are managed as flexible protected areas, in which human interventions such as agriculture and forestry are commonly recognized. They have thus been placed in the fifth IUCN category – 'protected landscape' – rather than the second.

Most western European national parks were created during the 1960s and 1970s. For example, France set up its first areas of this type – Vanoise and Port Cros – in 1963, while Germany established the pioneering Bayerischer Wald National Park in 1970. The reasons behind the demarcation of the parks were different in each case, and often contingent on a combination of local and national contexts. The Bavarian Forest (*Bayerischer Wald*) National Park, for example, was created in the mountainous, forest area along the border between Germany and Czechoslovakia – a marginalized region neighbouring the Iron Curtain – in order to stimulate local economic growth through the development of tourism (Von Ruschkowski and Mayer 2011). The implications of this process are discussed further in Chapter 7.

The Soviet Union's first national parks were founded in the 1970s, having been presented as a model involving parallel landscape protection and tourism development. They provided a significant alternative to the 'zapovedniki' – large areas set aside for scientific ecological research, and promoted by Lenin in 1921. However, the process also had strong ethnic undertones; the first national park – Lahemaa in Estonia – was created at the initiative of local elites interested in the protection of 'native' nature. Neighbouring Latvia and Lithuania quickly followed suit, establishing additional national parks on their territories. Smurr (2008) argues that these areas were set up by the Baltic intellectual and scientific leadership not only for nature conservation purposes, but also in order to 'guard' their national cultural heritage from intense Slavo-Soviet immigration. When the Soviet regime collapsed in 1991, Baltic countries developed their national park systems further, while establishing a joint organization – the Association of Baltic National Parks – in order to facilitate the governance of such areas (Kaltenborn et al. 2002).

Just like the Chinese example discussed above, many national parks in Europe (especially in central and eastern countries) were considered 'paper parks' until the 1990s, due to the lack of effective management practices. The last 20 years have seen not only the establishment of numerous new national parks, but also the revision and transformation of old ones. With its 27 member countries, the European Union (EU) – has played a crucial role in the transformation of national nature protection systems. Its vast Natura 2000 network incorporates many national parks, alongside other European landscapes that are evaluated and monitored according to a set of predetermined criteria.

Currently, most European national parks function as multi-purpose areas with an emphasis on promoting nature protection and sustainable livelihoods (see Figure 3.4). In countries where most of the land within the national parks is owned by the state – Germany, Sweden, Poland, Slovakia and the Czech Republic – these areas are managed as common public goods by one national authority with local executive branches; the system is often overseen or controlled by the ministries of environment or forestry (McCarthy et al. 2002). Still, there are major differences in the degree of centralization of management systems: in many cases, they are funded and administered at the regional or federal state level, even though national parks are designated nationally (Germany is a good example of this).

Many authors suggest that national parks are among the key income generators for local economies. For example, research undertaken by the National Trust in 2001 showed that three national parks in Wales (Snowdonia, Brecon Beacons and Pembrokeshire Coast) supported 12,000 jobs and generated a total income of 177 million British pounds. In Germany, one study estimated that approximately 50 million tourists visit the national parks each year, spend 2.1 billion euros in the process (Job 2008). Still, it is unclear whether this income can be directly attributed to the parks, considering that tourism was already developed in these areas well before their establishment.

It has also been emphasized that the branding of many European regions as 'national parks' has increased their popularity, making them highly attractive to second-home owners. However, the subsequent rise in property values has also decreased their affordability for local people, who are left with no other choice but to leave the area. Additional conflicts have been created by the trade-offs among conservation and other land use practices, such as agriculture, forestry, tourism and even industry.

Thus, the acceptance of the parks by communities depends largely on the strictness of the nature protection regimes, and the limitations stemming from them. This is illustrated by research undertaken in the aforementioned Bavarian Forest National Park. The park was initially managed as a recreational area, receiving widespread support from local communities (Hough 1988). However, a fierce storm in 1983 resulted in an extensive bark beetle infestation, affecting the locally highly valued spruce forests. Contrary to local expectations, the park authority chose not to intervene by removing the infected trees, leaving them in the 'hands of nature' instead. The rationale behind this decision was the expectation that the infestation might offer an opportunity for the natural regeneration of the extant spruce monoculture plantations. Thus, the park's governance changed from a pro-tourism management approach into a pro-conservation and rewilding model. As a result of this transformation, up to 95 per cent of the spruce forest in some parts of the park was destroyed by the bark beetle, leading to a complete change in the forest character of the landscape. The shift of practices has led to a shift of attitudes, because the majority of local residents have not accepted the new social and ecological settings of their homes (Von Ruschkowski and Mayer 2011).

Figure 3.1   Favelas and luxury housing developments are slowly encroaching upon the boundaries of the Tijuca National Park near Rio de Janeiro in Brazil

Source: Photo by author.

Figure 3.2   Table Mountain is an exception to the rule when it comes to African national parks, surrounded by a city, it is primarily an open access area

Source: Photo by Stefan Bouzarovski.

Figure 3.3   The establishment of Queensland's Lamington National Park was largely inspired by the Yellowstone model

Note: Visitors can enjoy the 180-metre length Tree Top Walk – the first of its kind in Australia – through the rainforest canopy.
Source: Photo by Stefan Bouzarovski.

Figure 3.4   The Vikos-Aoös National Park in northern Greece encompasses a complex landscape of historical human settlements and rare natural features

Source: Photo by author.

Concluding Remarks

Having reviewed the regimes and dynamics that have governed the establishment and management of protected areas – especially national parks – across the world, it is clear that a single global model for such territories does not exist in ideology or practice, despite the IUCN's standardized classification. This is mainly because, as has also been demonstrated by Chape et al. (2008), protected areas are most often the products of political interests and intentions among various elite groups operating at multiple scales. At the same time, the perceived 'value' of such spaces has been determined not only by natural attributes, but also by the economic and even the social viability of the objects of protection. The process has also favoured particular ecosystems (such as mountains), landscapes (i.e. forests) and species (apex carnivores such as orangutans, bears, lions) over others. Despite the creation of an extensive knowledge base about the importance of top carnivores and the hierarchical interdependence of species in ecosystems, such species have failed to receive consistent treatment in nature protection strategies due to shifts in social preferences. This is exemplified by the changing destiny of animals such as lions, leopards, baboons and birds of prey; once considered the 'vermin' of Africa, they were then protected, only for some of them to be consequently culled in a 'controlled' manner.

Although definitions of pristine and wild nature have also changed through time, the determination to conserve or even recreate controlled protected areas containing such characteristics has persisted. However, there have been major differences in the places, methods and times pertaining to the protection of these attributes, especially in terms of determining which parties should be responsible and benefit from them. The principles used to conceptualize and categorize protected areas have been a further source of polarization, especially with respect to the locally, culturally and socio-economically contingent nature-human relationships.

The recognition that, as a whole, protected areas have failed to achieve global biodiversity protection targets and decrease poverty rates has aided the emergence of more integrative approaches. Such thinking has acknowledged the connected and dynamic nature of society and nature in these territories, while arguing for the formulation of flexible and supranational frameworks to manage them. Yet it has not led the broader questioning of the rationales behind the rigid categorizations such as that of the UN; in many cases, rather, the application of supranational networks has generated the superposition of new categories and policies over old ones, creating a new layer of complication.

As I argued above, the 'national park' category is perhaps most emblematic of the multiple challenges and controversies faced by nature protection policies across the world. Despite the IUCN's efforts to place them under a single umbrella, the emergence of such areas has directly corresponded to the specific needs and interests of particular groups or nations. The driving forces behind their creation have differed across time, especially since the importance of the nature conservation component has been flexible and open for discussion.

The main reasons for creating national parks in the US, Canada, Australia during the late nineteenth and the early twentieth century can be found in the determination to conserve new-world pristine and 'natural' spaces for the recreational benefits of future generations, in addition to generating economic gains from the development of tourism. In these countries, the 'pure' nature of national parks has been portrayed as an antithesis to the polluted urban areas of the old world. However, the manner in which they define the public interest is problematic, in light of the fact that the creation of national parks has led to the displacement and marginalization of many local inhabitants, particularly traditional and indigenous groups and communities. The rights and interests of the local people were in most cases recognized in the 1990s, although there are still disputes over land-use and resources.

The mainstream recognition of the economic benefits brought about by tourism was a determining factor for the creation of the first national parks in several east Asian countries (particularly China and Japan), as well as New Zealand and Europe. In many cases, national parks have become popular recreational areas which are constantly engaged in improving tourist facilities without taking into account local contingencies involving land ownership, community participation and even nature conservation issues. As a result, such public 'natural' places have been transformed into fenced zoos or amusement parks.

In Africa, South America and South East Asia, the establishment of the first national parks was facilitated by the imperial powers' changing understanding of human-nature relationships. One of the main colonial rationales for the creation of such areas was contained in the protection of species and landscapes that did not hinder economic prosperity. Their borders were defined with little knowledge of, or respect for, the cultural and natural characteristics of the affected landscape. The strictness of management regimes and the level to which local people were given an opportunity to remain in their homes varied in accordance with the manner in which various colonial powers understood the purpose of such areas. Even though many countries maintained identical or similar governance regimes after gaining independence, pressure from advocacy organizations and human rights lobbies led to an increased recognition of local people's rights to resources and involvement in the management of the parks.

In some South American and African countries – particularly Argentina and Congo – national parks were created on the borders with neighbouring countries for geopolitical purposes, such as controlling migration flows. Even though a similar trend can be found in some Central European countries, the broader motives behind the construction of Europe's national park system were highly diverse. In Sweden and the former Soviet republics, the protection of national landscapes was seen as a useful tool for enhancing feelings of national identity. Many countries – Italy is a good early example – created their first parks so as to ensure the protection of specific 'charismatic' fauna. Overall, however, their underlying complexity and multiple functions has rendered all such areas fragile to stress and shock, and difficult to govern and adapt.

Chapter 4

Obstacles, Victims and Opportunists: Conceptualizing the Relationship between Local People and Nature Protection

Introduction

The previous chapters outlined some of the diverse spatial, cultural and economic processes, which have made the sustainable management of protected areas a key part of decision-making processes in developed and developing countries alike. However, effective and durable nature protection cannot take place without the balancing of conservation objectives with the socio-economic aspirations of local communities (Antrop 2001, Brockington et al. 2008, Hewett and Fletcher 2009). The active participation of local residents in the regulation of protected areas has increased the complexity of tasks faced by their management organizations, considering that the manner in which people perceive environmental quality and sustainability is influenced by, inter alia, socio-economic status, family ties and cultural affiliations (Wallner et al. 2007, Petrosillo et al. 2007). In part, this is because the socio-demographic structure of local communities may affect the acceptance of management strategies in a particular protected area, including the development of sustainable tourism and organic agriculture (Trakolis 2001).

Both the Third and Fourth World Congresses on National Parks, held respectively in Bali 1982 and Caracas in 1992, recognized the importance and urgency of improved local participation in the governance of protected areas. These events supported a conservation agenda that prioritized the sustainable development of protected areas, by acknowledging the rights of 'indigenous' and 'traditional' communities in their governance (Brandon et al. 1998). The Global Biodiversity Strategy and other policy frameworks that emerged from the 1992 United Nations Conference on Environment and Development in Rio de Janeiro (UNCED, the 'Earth Summit') operationalized some of these principles by emphasizing the importance of management tools such as buffer zones, community-based conservation, biosphere reserves and bioregional management (Zimmerer et al. 2004).

Academic and policy-based scholarship has tended to conceptualize nature protection regimes as belonging to either a classical nature-orientated, a neo-populist human-centred, or a neoliberal, market-focused approach (Blaikie and Jeanrenaud 1997). Much has been written about the ways in which these three paradigms relate to wider political, social and spatial dynamics, as well as their embeddedness in particular historical paths and spatial circumstances (Mulder and Coppolillo 2005, Bajracharya and Dahal 2008, Mathur and Sinha 2008). Researchers have paid comparatively less attention, however, to the underlying ideologies and policies that have guided the processes through which local people and communities are implicated in these modes of nature conservation. With the exception of particular geographies (e.g. Africa and Latin America), simplistic views that see local populations in a highly instrumentalized and reductionist manner are common in the literature on environmental politics and management (for a discussion, also see Spinage 1998, Castro and Nielson 2004). Both the extent to which the different ways of governing protected nature have been connected in time and space, and the role of local people in this process, have been marginalized in the theoretical corpus of relevant academic disciplines.

The aim of this chapter, therefore, is to provide a critical overview of the different ways in which local populations have been represented and treated by academic discourses on, and analyses of, nature protection. Having surveyed more than 60 academic and policy-orientated contributions – a small fraction of the vast literature on the topic – I trace the historical evolution and main features of various theorizations of the people-protected areas relationship, with the aim of conceptualizing mainstream perceptions of local people's needs, aspirations, and public participation issues. The chapter also briefly discusses the broad-level environmental governance implications of different academic and policy understandings of the role of local people in protected areas, including co-governance, and adaptive and resilience management. By highlighting the diverse pathways through which such policies affect, and are shaped by, local populations' perceptions, attitudes and opinions, I wish to undermine the binary between developed and developing countries that pervades the literature, emphasizing the role of place-specific circumstances instead.

My underlying motivation and argument is that the residents of protected areas have often been disempowered by the conceptualization of their relationship with nature conservation. Initial representations of the inhabitants of protected areas – some of which still linger today – saw them as obstacles towards effective conservation, who must be either physically removed from protected areas, or face limitations on economic activities (Terborgh 2004). In response to the social and environmental injustices that arose as a result of the policies stemming from this approach, a discourse of local populations as 'victims' appeared in the relevant scholarship on the topic (Colchester 1997, Agrawal and Redford 2009, Dowie 2009). This so-called 'neo-populist' discourse has been criticized in its attempt to often mythologize the residents of protected areas, by focusing on their ancestral ties to the land, folklore, and the alleged possession of traditional knowledge. In doing so, I would argue, such thinking has also disenfranchised locals by denying the pragmatic basis for their actions. Still, later attempts to move beyond the victim-obstacle binary have often subjected local populations to further marginalization by framing them within top-down co-management dynamics, in order to meet bureaucratic goals and organizational aims (Baird 1999, 2000, Mulder and Coppolillo 2005, Brockington et al. 2008, Duffy 2010, Büscher and Dressler 2012). As a result, the voice of locals in protected areas is rarely heard, which often forces them to either represent themselves as objects of tourist consumption, on the one hand, or face impoverishment and political marginalization, on the other.

My exploration of these issues is framed through three sets of perspectives, where the role of local people in nature protection is seen through the lens of, respectively, obstacles, victims and opportunists. Being aware of the dangers that generalization brings, however, I would like to emphasize that this is merely a stylized triptych, which is meant to provide a starting point for articulating a political ecology-based critique of mainstream views of local population involvement in the management of protected areas. The chapter thus serves as a starting point for a broader exploration of the need for alternative community-based conservation approaches – in addition to the development of flexible and resilient socio-ecological systems – which is presented in the empirical explorations that follow in the remainder of the book.

In light of the numerous controversies surrounding the delineation of nativity and local residence (Mulder and Coppolillo 2005) I equate the syntagm 'local people' with the standard definition of the term 'resident' provided by the Oxford English dictionary: 'a person who lives somewhere permanently or on a long-term basis'. The limitations and advantages of this definition, and its operationalization in my empirical investigations, will be discussed in the chapters that follow.

Local People as Obstacles towards Nature Protection: Fortress Conservation and its Discontents

People living within and around protected areas transform the landscape features by undertaking various land-use practices. Their daily activities affect the ecosystem functions and the biodiversity abundance. Hence, conservationists and natural scientists have propagated the argument that specially protected places valued for their abundance of species and geomorphological features must be protected from the surrounding human impacts. The blueprint for such nature conservation practices was taken from early attempts to protect a particular area or species throughout the world. This model, known as 'fortress conservation', is based on the North American idea of a national park and the European understanding of managed nature reserves in the nineteenth century (Wells and Brandon 1992, Neumann 1998). Although such geographies were key for the theoretical development of this model, North America first tried its practical application with the creation of the large national parks such as Yellowstone and Yosemite (see Figure 4.1). Soon afterwards, it was rolled out across the world with special upsurge in the colonial countries in Africa, Asia and South America.

The successful transfer of this approach from the United States to many other countries across the world meant that local populations started to be perceived

Figure 4.1   Fortress conservation in action; a woodland restoration project in Yosemite imagines the 'Miwok people' as part of the protected landscape, rather than a distinctive social and political entity

Source: Photo by author.

as obstacles towards nature protection. The neo-Malthusian argument that the increasing human population around and within the protected areas leads to depletion of the valued conserved resources has been very often used as supporting evidence for the exclusion of people from these areas. As a result, relocation and displacement programmes in the name of nature conservation became common practice, especially across Asia and Africa during the 1970s (Brockington and Igoe 2006). The subsequent emergence of more inclusive and community-based approaches – which emphasized the need to include local people in the governance of protected areas (Hutton et al. 2005) – was seen by 'fortress conservationists' as an effort to promote a neo-populist approach that threatens the survival of protected areas and species. They argued that community-based conservation fails to provide the promised win-win situation, including well-preserved natural areas and the improved wellbeing of local people. Hence, the articulation of conservation work grounded in science (Wilshusen et al. 2002) was seen as more than necessary in order to save natural habitats from 'doomsday'.

The 1990s saw a rising backlash against community-based conservation, which Hutton et al. (2005) terms 'back to the barriers'. In their words, such thinking 'reasserts that biodiversity can only be conserved in areas free of all human influence' (347), with the exception of science and limited ecotourism. Also, 'back to the barriers' thinking employed conservation science 'to identify the best areas for reserves', which were then seen as playing the role of 'filling gaps' in a wider 'system or network of protected areas' (ibid.). However, establishing a wider apparatus of this kind also created a situation where all available resources are focused on expanding and completing the existing system, resulting in 'landscape-scale conservation'. It holds, therefore, that:

> ... such protected areas are best managed by centralized authorities, closely policing marked boundaries and applying appropriate sanctions on those who violate rules and borders, particularly those who enter reserves to reach the other side or to hunt, fish, gather plant materials or graze animals. (Hutton et al. 2005: 347)

The new wave of the 'fortress conservation' ideology was disseminated by a number of publications by authors such as Kramer (1997) and Terborgh (2004). Despite the accusations, many conservationists firmly believe in the principles of this cause:

> No apology should be required for adhering to the expected definition of a (national) park as a haven for nature where people, except for visitors, staff, and concessionaires, are excluded. To advocate anything else for developing countries, simply because they are poor (one hopes, a temporary condition) is to advocate a double standard, something we find deplorable. (Terborgh and van Schaik 2002: 6)

Arguing along the same lines, Diamond (1995) has emphasized the devastating impact of indigenous people on the forest cover and its constituent species on Easter Island. A similar claim is advanced by Lovejoy (2002), who cites abundant evidence about changes in biodiversity within islands in the Caribbean, the South Pacific and Hawaii. In such areas, indigenous people eliminated numerous endemic species long before the arrival of European settlers. While Miranda and LaPalme (1997: 134) argue that user rights exercised by local people can sometimes have devastating effects on local ecosystems, Terborgh et al. (1999) have found indications of low density prey populations with an almost eliminated ecosystem function, even where levels of hunting are sustainable. Peres (2000) notes that this is also the case with traditional and indigenous populations – a claim further elaborated by Robinson and Bennett (2000), who show that large game animals are highly valued as prey animals and as such are strongly depleted in places where human population densities are over one person per km^2. These habitats commonly include tropical forests across the globe.

The view of local people as obstacles towards effective nature protection is also common in theorizations of management and policy issues. A large part of the literature in this vein has insisted that 'sustainable development other than ecotourism is incompatible with nature conservation', since 'humans and animals do not mix well' (Terborgh and van Schaik 2002: 6). As far as management approaches are concerned, Wells and Brandon (1993) insist that integrated projects with combined conservation and economic development objectives may fail to guarantee biodiversity. Similarly, Spinage (1998) dismisses participatory conservation, claiming that its supporters sacrifice 'conservation to political correctness' (cited in Dowie 2009: 87). Salafsky and Wollenberg (2000) suggest that core protected zones with no consumptive use of biological resources should still be central to protected area management.

Advocates of the 'local people as obstacles towards nature protection' view have often focused on the dynamics of knowledge and the flow of information in the governance of nature conservation, pointing out that the involvement of the residents of protected areas in their management is problematic, as a result of the lack of knowledge about the workings of participatory democracy and the functioning of ecosystems. Thus, Terborgh (2004) emphasizes that only science and institutions in the traditional sense can 'save nature'. Scientists and practitioners working within this vein tend to view Integrated Conservation and Development Projects (IDCPs) – which are aimed at, *inter alia*, improving the co-operation process between people and protected area authorities – as 'rural development programmes that confer only incidental benefits for nature conservation' (van Schaik and Rijksen 2002: 18). Regarding local people's attitudes towards protected areas, Carrus et al. (2005) argue that negative relationships between local identity and nature conservation can be expected when a protected area is likely to create a situation in which the local residents' individual interests clash with some general collective interest imposed by extra-local authorities. However, protected areas rich in biodiversity have been valued as globally irreplaceable, and as such have become 'transnationalized spaces of high biodiversity value' (Ferguson 2006 cited in Igoe and Brockington 2007: 441).

**Locating the Human Victims of Protected Nature:
The Advent of 'Neo-populism'**

The 'fortress conservation' model has been frequently discussed by, and criticized from, human rights and environmental justice standpoints; numerous activists and scholars have pointed to its detrimental consequences on the welfare of local populations (Colchester 1997, Ghimire and Pimbert 1997a, Adams 2003). As noted previously in this chapter, the new wave of 'community-based conservation' initiatives during the 1990s provided a diverse range of approaches, including community wildlife management, collaborative or co-management, community-based natural resource management, state/community co-management and integrated conservation and development programmes – ICDPs (Barrow and Murphree 2001).

The emergence of a wide range of co-management and participatory practices in recent years – including the ICDPs mentioned above – stems from the rise of political and scholarly attempts to alleviate and prevent the marginalization of indigenous and traditional peoples resulting from the 'fortress conservation' model. In one of the key papers describing the history of participatory conservation, Hutton et al. (2005) pointed out that 'in the 1980s, decentralized, community-based approaches to biodiversity conservation and natural resource management began to spread rapidly, especially in southern Africa' (341). They emphasize that:

> By the 1990s, the dominant narrative of fortress conservation no longer enjoyed hegemony, either in Africa or globally. It had progressively been challenged by a new community conservation narrative which stressed the need not to exclude local people, either physically from protected areas or politically from the conservation policy process, but to ensure their participation. (ibid. 342)

Blaikie and Jeanrenaud (1997) place many of these initiatives under the aegis of a 'neo-populist approach' whose main emphasis is on questions of environmental and social justice. Colchester (1997) argues that protected areas have imposed elite visions of land use, resulting in the alienation of common lands. He adds that 'national parks established on indigenous lands have denied local rights to resources', as a result of which local people have been turned from 'hunters' and 'cultivators' into 'poachers' and 'squatters' practically overnight (13).

Many research and policy contributions within this vein were motivated by broader theoretical thinking, which challenges the distinctiveness of human-nature relationships and the current nature conservation practices and tendencies (see, for example, Goldman 1998). One of the broadly challenged nature conservation perspective is the obsession to conserve, create or recreate the 'wilderness'. Thus, Cronon (1995) criticizes the 'Western myth of wilderness', which is predicated upon the claim that nature can somehow be left untouched by human presence. When talking about biodiversity conservation, he claims that the prime objective of early American national parks was to protect spectacular scenery, rather than biological resources. Within these initiatives, 'far from being the one place on earth that stands apart from humanity' and the 'contaminating taint of civilization', wilderness is precisely a product of human cultures within particular times of human history, and as a result can 'hardly be contaminated by the very stuff of which it is made' (Cronon 1995: 97).

Even though evictions and relocations of local people from protected areas are officially not allowed in many countries, issues regarding the consequences of past and possible future actions of this kind are still highly topical in the literature (Brockington and Igoe 2006, Brockington et al. 2008, Agrawal and Redford 2009). In stressing the negative effects of Western-style conservation, Duffy (2010) argues that local people continue to be seen as a nuisance in protected areas, suffering restrictions on their way of life and even extrajudicial killings in the name of nature conservation. Systematic and detailed evidence about the consequences of eviction and limited opportunities for the use of natural resources – including impoverishment and political exclusion – has now been gathered and interrogated for a wide variety of geographical contexts and cases (Ghimire and Pimbert 1997a, Hulme and Murphree 2001, Rao et al. 2002, McLean and Straede 2003).

Dowie's (2009) investigation of the reasons for the economic marginalization of the resident populations of protected areas emphasizes that this dynamic is rooted in the fact that 'indigenous people are moved into the lowest end of the money economy, where they tend to be permanently indentured as park rangers (never wardens), porters, waiters, harvesters, or, if they manage to learn a European language, eco tour guides' (xxvi). The implication of such processes, he argues, is that 'conservation', becomes 'development' resulting in assimilation of 'native communities' into 'national cultures' (ibid.). His highly critical account of perspectives that see local people as obstacles towards nature conservation underlines the role of 'population growth, erosion of culture, market pressures, and the misuse of destruction technologies' (111) in the rise of practices such as poaching and squatting.

It is frequently pointed out that the 'fortress conservation' model has led to the relegation of local people to a reduced socio-economic role as 'parts of the landscape', involving highly restricted agricultural and/or forestry activities, but with an encouragement for tourism development. Balakrishnan and Ndhlovu (1992) point to the widespread resentment against safari hunting that has resulted from the Yellowstone nature protection model, which is seen to be giving outsiders opportunities for the utilization of wildlife resources that locals often have no access to. Arguing along similar lines, Barrow and Murphree (2001) stress that even the institutions that were effective in the protection and management of nature undermined local communities in Africa in the 1990s. At the same time, Ghimire and Pimbert (1997) underline that:

> … most protected area management plans and evaluation reports avoid even referring to structural issues such as land reform, income distribution, decentralization of power, social mobilization, as well as local rights and sovereignty over resources, without which sustainable management and more socially oriented use of natural resources in rural areas is mere illusion. (32)

These authors also emphasize that 'conservationists and protected area managers have to take into consideration the role of human beings', since they are 'one of the species inhabiting ecosystems' (ibid. 5). Indeed, a survey of 37 projects aimed at simultaneous biodiversity conservation and poverty alleviation has come across systematic evidence favouring the mutual synchronization of these two objectives (Agrawal and Redford 2006). Colchester's (2004) work supports this argument, by claiming that new conservation models should respect the rights of indigenous peoples and other bearers of 'traditional knowledge'. He also argues that 'conservation policies emerged at a time of fierce prejudice against indigenous peoples' helping build the global acceptance of a model of 'colonial conservation' (ibid.). In his view, this process has 'caused, and continues to cause, widespread human suffering and resentment':

> The history of indigenous peoples' relations to protected areas can be seen as one of social exclusion and marginalization. Having once been independent nations within their own territories, indigenous peoples have been pushed out of their lands, which have been expropriated by government agencies in the name of conservation … International law recognizes that indigenous peoples have rights to own, manage and control their lands and conservation policies have accepted this in principle. The challenge is to allow indigenous peoples to move back into control of their lands. (Colchester 2004: 145–152)

Such claims are in line with Article 8(j) of the 1992 Convention on Biological Diversity – adopted at the Rio Summit – which states that:

> … each signatory nation, subject to its national legislation will respect, preserve and maintain knowledge, innovation and practices of indigenous and local communities embodying traditional lifestyles relevant to the conservation and sustainable use of biological diversity. (149)

However, Dowie (2009) is pessimistic about the ability of co-management approaches to prevent the social exclusion of local populations, since they rarely encourage local stewardship and are not immune to the above problems. He points out that evictions from protected areas in states where such practices are illegal are often veiled behind co-management projects and presented as voluntary relocations. Moreover, it appears that not only traditional conservationists are against joint conservation and development projects, since some political advocates of local communities' rights to use and manage natural resources have also been unsatisfied by co-management practices (de Castro et al. 2006). Criticism of top-down management approaches has also been voiced from a theoretical perspective; in putting forward the idea of resilient ecosystem governance, Folke et al. (2002) underline that rigid environmental governance can erode resilience and promote the collapse of ecological-social systems.

Conservation models based on the neo-populist paradigm place a major emphasis on the involvement of 'indigenous' people with their 'traditional' knowledge in the management of protected areas (Blaikie and Jeanrenaud 1997). However, Adams and Hulme (2001) note that the restrictions that are supposed to keep such populations 'traditional' represent a 'naïve view' that does not provide an effective policy for the conservation of biodiversity. A powerful empirical illustration of this argument has been provided by Ghimire and Pimbert (1997), who underline that in certain states of Brazil, the government has decided to allow only 'traditional' populations to remain inside national parks, threatening many local communities along the 'traditional-non-traditional populations gradient' (de Castro et al. 2006). It is also claimed that labelling certain populations as 'traditional' inherently carries with it the danger of commodifying their customs and practices for tourist consumption, while symbolically undermining their role in contemporary political processes (Mulder and Coppolillo 2005).

Despite being frequently framed under the aegis of the developing/developed country binary, there is ample evidence that issues of local participation and the political challenges associated with the establishment of co-management practices are also present in developed country contexts. Papageorgiou and Vogiatzakis (2006) point out that 'nature vs. humans' issues can also be found in protected areas in Europe, underlining the specific problems regarding the implementation of Natura 2000, the European Union's network of protected areas. Citing the example of Greece, they argue that the superimposition of Natura 2000 upon the existing system of protected areas resulted in the duplication of administrative efforts and related legislation, making the overall management of protected areas 'complex, confusing and fragmented' which further complicates the participative management practices (ibid. 476). Young et al. (2007) stress that stakeholder participation in Central and Eastern Europe has also been rather limited to date, a conclusion that has motivated Papageorgiou and Vogiatzakis (2006) to suggest that incentives should be given to local communities in Europe in order to benefit from protected areas. In her book about the Adirondack protected area in the United States, Knott (1998) also stresses that some voices still have not been heard, especially those of marginalized local Native Americans. Like the other authors working in this vein, she underlines the need for improved participative management in the nature protection process.

There has been a lot of controversy over the question of who should benefit from natural resource use incentives in protected areas. The idea that the focus of such policies should be on 'indigenous' or 'native' people has frequently come under fire. Mulder and Coppolillo (2005), for example, think that labelling local people in protected areas as 'traditional' or 'native' is 'awkward'. They cite two reasons for this, the first being that the label 'traditional' (or 'native') necessarily restricts focus to communities that are 'indigenous'. They see this as a problematic category, 'partly because it is rarely known how long a group of people have inhabited an area' and 'whether they were the first to do so' (Mulder and Coppolillo 2005: xv). According to them, also of relevance here is the fact that the term 'indigenous' may be associated with the idea of a 'harmonious relationship between the human community and the environment, a view that is open to question' (ibid. xv). The second reason for these authors' claims resides in the fact that 'the principal threats to biodiversity come from habitat loss and overexploitation' (ibid.) which means that a range of groups 'from local fishermen to logging operators' may be responsible. By viewing such debates via a constructivist lens, Robbins (2004) argues that any categorization – including the grouping of local people under labels such as 'indigenous' or 'native' – may adequately capture some commonalities in the pattern of reality, but is ultimately no more accurate than other possible classifications.

The environmental governance implications of the 'indigenous' vs. 'non-indigenous' binary have been explored across a number of scholarly contributions. Special attention has been paid to the controversies related to the recognition of Maasai rights over Tanzania's natural resources in the 1990s (Neves and Igoe 2012). Even though the Maasai are relative newcomers to the area, this development was followed by the establishment of 'indigenous' Maasai areas in Tanzania centring on claims of identity based on ethnicity. While democratization has provided the necessary political space for the articulation of such declarations, the wealth of economic opportunities – development funds, workshops, overseas tours, and so on – offered by international donors has led to bitter competition among them. One of most contested questions relates to the legitimacy of place identity, as there are other (non-Maasai) pastoralists in northern Tanzania. Hodgson (2002) cites the example of a workshop participant who stated that 'Some people just put red clothes and call themselves pastoralists' (1090) when referring to the Maa-speaking Arusha people who have recently reclaimed their Maasai heritage in order to accommodate donor preferences to help the 'Maasai'. Igoe's (2010) research on the topic has found that these arrangements have marginalized people who are less 'culturally distinct' than their Maasai neighbors, and thus 'aesthetically at odds with the imagined natures that western tourists pay to view'. The perception of Maasai people as being 'close to nature', he argues, has led to longstanding conflicts between local communities and government conservation authorities (ibid.).

European protected areas are not immune to such situations. The vast region of Lapland in Sweden – including the Padjelanta, Sarek, Stora-Sjöfallet and Muddus national parks as well as several other protected areas – has been plagued with a range of conflicts relating to nature protection. They involve the 'indigenous' Sami people, who have been living in the area for around 4,000 to 5,000 years and have been granted greater rights in using natural resources, versus the non-Sami population which is considered 'less' local – since its historical record of habitation dates back less years – and therefore entitled to lower levels of resource use (UNESCO 2006). The Saami are legally permitted to herd reindeer and to hunt and fish over 16,000,000 ha of both private and state land. Some of the most significant clashes exist between Sami herders and the owners of grazing areas (mostly private forest or farmland); members of the latter group on compensation for the damage caused by browsing reindeer (Swedish Institute 1990).

Some protected areas in western European states, including the UK, face conflicts between the antecedent rural population – which often tends to include low-income households that are that highly dependent on local natural resources – on the one hand, and recently-settled second-home owners, who are usually urban middle class dwellers requiring 'pure' nature, tranquillity, and strict limits on the use of natural resources, on the other (see Figure 4.2). Southern European national parks are often enveloped in problems relating to the construction of tourism-related facilities, whose owners may buy and build on agricultural land without planning permission (IUCN 2010). In turn, this creates infrastructural difficulties

Figure 4.2   The attractiveness of the United Kingdom's Lake District National Park for tourism and second-home construction has led to 12:1 house-price-to-income ratios in the area

Source: Photo by author.

and pollution, which disproportionately affect the residential populations of protected areas. Arguing along the same lines, Piermattei (2013) has found that the creation of the Monti Sibillini National Park in Italy contributed to the stratification of local power relations by promoting wilderness and eco-tourism-orientated policies, while politically marginalizing the interests of small scale local farmers and shepherds. Despite the apparent lack of forced displacement or the denial of land-use rights, the park generated its own distribution of power, which disables specific groups from participating in policy and decision making at the grassroots level.

Turning Protected Area Inhabitants into 'Opportunists' and 'Natives': Neoliberalism Takes Over

Despite its beneficial effect on human welfare, numerous experts have insisted that community-based conservation has been unable to provide an effective framework for the management of social and economic pressures on protected areas (van Schaik and Rijksen 2002). The 'neo-populist' approach that encompasses this policy has been critiqued for its inability to protect biodiversity, physical landscapes and 'wilderness' in nature reserves (Terborgh and van Schaik 2002). Its failure has often been attributed to the fact that some local people are uninterested, unwilling or unable to participate in nature conservation.

Moreover, it has been argued that the relevant academic literature has tended to conceptualize local residents and communities as either victims of, or obstacles towards, the sustainable management of national parks, as a result of their purportedly entrenched negative attitude towards nature conservation and protection. But there is insufficient empirical evidence to support the veracity of such claims across different regulatory and geographical settings (Lawrence 2008).

The combined consequence of such arguments and broader socio-economic developments in the 1980s and 1990s has been the advent of a neo-liberal approach towards nature protection. Its central tenet is a staunch reliance on the powers of free market allocation, embodied in the suggestion that natural resource management should be taken away from the hands of the state. At the same time, the advocates of this paradigm recommend removing the incentives that encourage the non-sustainable use of resources, while stimulating the 'internalization of environmental costs' (Blaikie and Jeanrenaud 1997: 64). Even though the neoliberal view neither criticizes nor romanticizes local people – unlike, respectively, the classic and neo-populist approaches described above – it has nevertheless played an important role in the re-distribution of power and resources, often resulting in the further impoverishment of marginalized groups (Dey 1997).

At this point, it is worth noting the extensive body of literature on the practices and discontents associated with neoliberal conservation. Much of this research uses a critical lens to link the constitutive practices of neoliberal economics – privatization, marketization, liberalization – with nature protection and social justice. As pointed out by Igoe et al. (2010) conservationists who require financing from external sponsors 'are increasingly under pressure to show the economic advantages of their conservation goals' (ibid. 487), as a result of operating within a capitalist framework in which 'the allocation of funds is directly related to associated potential returns on investment' (ibid.). This creates an increasing collusion between the notions of 'conservationist action', on the one hand, and 'capitalist reality' on the other, whereby the suggestion that the one is beneficial to the other becomes taken for granted. The systematic and extensive promotion of such approaches leads to the emergence of a hegemonic idea, whereby capitalist frameworks are represented as 'the only feasible view' (ibid.) of the effective setting and application of conservation goals. However:

> The hegemonic nature of these claims, and the ideological context from which they are derived, present major obstacles to democratic and reflexive discussions of our most pressing socio-environmental problems. Capitalism's destructive-extractive relationships with the environment are unlikely to be challenged unless we are able to understand the fundamental contradictions between capitalism's need to expand exponentially vis-à-vis the capacity of ecosystems to withstand

> and absorb the disturbances and stresses that this exponential growth entails.
> (Igoe et al. 2010: 487)

Management tools developed under the aegis of neoliberal conservation have attempted to provide economic motives for the sustainable management of nature reserves and parks. They chime in with a broader effort to introduce capitalist-based market transactions 'to the way in which more-than-human nature has been conceived, controlled, distributed, managed and produced' (Heynen and Robbins 2005: 6). This has been remarkable both in the case of common property resources, which have long resisted enclosure due to their fugitive nature, as well as in the privatization of urban nature, where stark environmental inequalities have been produced through new systems of property and governance. Furthermore, these authors point out that:

> Overcoming the socionatural difficulty of enclosing water, air and wildlife, emerging property and governance regimes have devastated public ecologies and quashed resistance to enclosure in both urban and rural contexts. Market systems are being extended to fish, carbon and water, with scarcities becoming an increasing result rather than cause of this re-institutionalization. (Heynen and Robbins 2005: 6)

Practical experience relating to the implementation of community-based approaches has provided some important lessons for the latest generation of biodiversity conservation programmes, including those concerned with poverty alleviation, as well as those working at ecosystem and landscape scales. More recent perspectives have started to incorporate elements of adaptive management, new partnership models with stakeholders, and the vertical integration of site-level work with policy initiatives and institutional development. For example, Phillips (2004) stresses that there is enough evidence to promote the idea of Community Conservation Areas (CCAs) as a useful nature protection tool. He outlines some of the advantages of 'sacred groves' in Africa, 'tapu' areas in the South Pacific, 'hemas' reserves in pastoral communities of west Asia, 'indigenous protected areas' in Australia and 'regional natural parks' in France. It should be emphasized that all spaces of this kind are classified as 'protected landscapes or seascapes' (category V) by the International Union for the Conservation of Nature, and as such are more socially inclusive and sensitive to local needs.

While such cases refer to a specific type of protected area, examples of successful local participation in broader practices of environmental management and protection at the national scale can also be found under the auspices of Community-Based Natural Resource Management (CBNRM). In Philippine national parks, local people are, to a large extent, the de facto day-to-day managers of biological resources, given their physical proximity to them, coupled by an intimate knowledge of fauna, flora, and habitats. The role of the government is limited due to the remoteness of protected areas, and the persistent shortage of funds. In addition, people maintain traditional community-based systems for controlling and monitoring access to certain sites and resource use, while community leaders and people regularly discuss the availability and quality of natural resources. The development of participatory monitoring methods has benefited from the existence of such informal systems. In Tanzania, the shift in paradigm gained legitimacy in 1998 with the introduction of a wildlife policy that brought about the creation of 'wildlife management areas'. These spaces allowed local communities, including women – a user-group traditionally excluded from decision-making processes – to receive tangible income from setting aside land in their village for the purpose of preservation and protection of wildlife.

However, some scholars and practitioners suggest that CBNRM is experiencing a crisis of identity and purpose, with even the most positive examples experiencing only fleeting success due to major deficiencies (Büscher and Dressler 2012). Thus, Kellert et al. (2000) have uncovered a number of problems with the implementation of CBNRM across the world, 'despite sincere attempts and some success' (705). They note that CBNRM has rarely resulted in a more equitable distribution of power and economic benefits within Nepal and Kenya in particular. In such countries, the approach has often failed to protect biological diversity, stimulate the sustainable use of resources, reduce conflict, and increase the consideration of traditional or modern environmental knowledge. Conversely, the implementation of CBNRM in the North American context has been more successful. These authors argue that the observed differences can be attributed to institutional, environmental, and organizational factors. Similar challenges have been identified in the case of aquatic resource co-management (Baird 2000), even though authors in this field emphasize that it is possible to achieve real benefits over a short period of time, provided that local people are allowed to participate and planning is co-ordinated well. These arguments are supported by, *inter alia*, Hogan (1997) and Baird (1999), who provide positive examples regarding the implementation of aquatic resource co-management in Laos and Cambodia.

Concluding Remarks

The last two decades have seen the rise of flexible approaches towards nature conservation management of protected areas across the world, marked by a deeper understanding of ecosystems dynamics. Initially, the establishment and operation of protected nature was guided by biodiversity conservation concerns, encompassing ecological principles based on population regulation, island biogeography and metapopulation dynamics (Ale and Howe 2010). With cultural and social conservation starting to play an increasingly important role in this context, however, it has been frequently pointed out that protected area governance often clashes with the ecological approaches that are meant to inform it (Jentoft et al. 2007). While Ale and Howe (2010) attribute the emergence of this situation to the fact that protected area managers may expect too much from paradigmatic blueprints, Bawa et al. (2004) stress the importance of the practice of conservation rather than its subject. They highlight the need for creating a sufficiently flexible and open institutional framework for nature protection, capable of stimulating the flow of information and taking into account human concerns (Duffy 2010). Such thinking is often motivated by the conclusions of the significant body of political ecology scholarship, which insists that conservation initiatives must rely on historical and geographical specificities as starting points (Neumann 1998).

In this chapter, therefore, I have highlighted three broad-level conceptualizations of the relationship between local people and nature protection, which emerged successively during the twentieth century. The three approaches correspond with the principles of fortress conservation, 'neo-populism' and neoliberalism. Although the victims-obstacles-opportunists triad has not been explicitly theorized as such in the relevant academic and policy literatures, academic work in this field has often highlighted the ideologies and politics that drive environmental governance in protected areas, as well as the different ways in which nature protection regimes have been implicated in shaping the livelihoods and everyday activities of local people.

At the moment, the various policy embodiments of all three perspectives – including the Yellowstone model, CCAs and CBNRM – exist simultaneously across different parts of the world. This can be partly attributed to the lack of a consensus about their environmental effectiveness and social justice implications; each view has its critics and discontents. For example, the conceptualization of local people as 'victims' emerged largely as a reaction to the failures of state-run exclusionary conservation. The policy embodiment of this perspective has attempted to develop governance practices that are more inclusive and sensitive to the needs of the inhabitants of protected areas. But although Blaikie and Jeanrenaud (1997) point out that very few conservationists would dare to voice the 'fortress mentality' today, the views of many of its adherents – such as Salafsky and Wollenberg (2000), van Schaik and Rijksen (2002), and Terborgh (2004) – are still widely respected. This cacophony of voices, I would argue, has created a hybrid geography of nature protection regimes with respect to the rights and roles of local people.

In this context, it is worth noting Brockington's (2004) comments on the relationship between social justice and nature protection. He notes the existence of 'a persistent error in social science writings about conservation, which is also particularly prevalent in sustainable development writings generally; that unjust conservation does not work' (84). He criticizes this 'oft repeated mantra' for conflating 'social justice as end in itself and as a pragmatic means of achieving conservation goals' (ibid.), while emphasizing that 'oppression and injustice have been sustained repeatedly for centuries, and continue to be sustained'. The implication of 'asserting that inequitable states of affairs cannot last', he contends, is that we are stopped 'from asking when they might be changed or how' (Brockington 2004: 85). He, therefore, sees the primary question as one which focuses on the kinds of circumstances under which 'oppression is effectively maintained' as well as the conditions in which 'the rural poor can effectively oppose unjust conservation' (ibid. 85). Therefore:

> We must find the common ground, the innovative spirit, and the political will to help rural people become effective partners in conservation that provides them with equitable benefits while effectively protecting biodiversity. (Brockington 2004: 85)

Thus, one of the shortcomings of the obstacles–victims–opportunists triad is that it may marginalize a fourth argument: that local people are vitally important, but deliberately excluded. It is therefore worth emphasizing work by authors who try not to homogenize local communities, but instead point out who benefits and who is disadvantaged by nature protection (for example Brockington et al. 2008). Indeed, much of my discussion in subsequent chapters is developed in this vein.

This review has also demonstrated that it is difficult to place policy issues stemming from the nature protection-local people nexus within the spatial dichotomy of developed and developing world countries. While much of the literature tends to conceptualize questions of local participation and social justice as being on either one of the sides of this binary, my impression is that the assumptions that many authors use when generalizing their results from local case studies to developed or developing country contexts are largely based on weak evidence. This is because the experience of local people across the world indicates that threats of eviction and curtailment of natural resource use are still common on both sides of the divide between the global north and south. In addition, developed and developing countries alike contain attempts to commodify and objectify the human inhabitants of protected areas, by relegating them to the politically safe role of 'traditional' elements within a landscape destined for tourist consumption.

The question as to whether it is possible to move beyond the obstacles–victims–opportunists triptych, however, remains open. Although the importance of local community engagement in protected area management has been widely recognized and analysed in the academic literature – having been followed by a rising involvement of local populations in rural development decisions – there is limited empirical evidence that 'community participation has become widespread practice or been effective in influencing the nature and scale of development' (Goodall and Stabler 2000: 63).

Looking at the disciplinary backgrounds of the authors that I have cited quickly reveals that the three conceptualizations are embedded in different ideologies and epistemological cultures: while a disproportionate number of scientists and biodiversity specialists are in favour of fortress conservation, the most vocal advocates of the neo-populist approach can be found among anthropologists and social geographers; at the same time, neoliberal economists tend to support the view of local people as opportunists.

Indeed, the segregation of disciplinary backgrounds in relation to different conservation ideologies has been highlighted by a number of authors. Most notably, Agrawal and Ostrom (2006) claim that conservation biology and its 'central concern, biodiversity' have been largely ignored by political scientists. As evidence for their argument, they cite indicators and an anecdote concerning 'publication trends and faculty hiring in political science' demonstrating that only one article focusing on biodiversity conservation has been published in the political science journals with top impact factors scores, out of 2,000 published papers. They have also been unable to find a single academic within the 'top five political science departments in the United States (arguably, Harvard, Princeton, Stanford, Yale, and Chicago)' who has listed biodiversity as one of their areas of research interest. 'These facts', Agrawal and Ostrom (2006) contend 'should rightly generate pessimism' about the past and future ability of political scientists to enter a dialogue with conservation biology, as they demonstrate that 'political scientists and their discipline value work on biodiversity at best only to a limited extent' (681). However, to the extent that political scientists consider conservation biology and biodiversity, 'they must do so by working against the disciplinary incentive structures that reward research and teaching' (ibid.).

Developing a sufficiently critical and comprehensive understanding of the advantages and shortcomings of the different perspectives – while perhaps moving to a fourth, more effective and just solution – will require a significantly higher degree of inter-disciplinary integration, accompanied by a more explicit focus on local geographical and temporal variations in the environmental governance challenges faced by the residents of protected areas.

Chapter 5

Nature Protection and Municipal Government During Socialism and After: Legacies, Challenges and Opportunities

This chapter provides an overview of the historical legacies, regulatory arrangements, policy regimes and spatial patterns associated with the governance of environmental issues and nature protection in Central and Eastern Europe. At the centre of the chapter lies an endeavour to move debates about nature protection beyond the relatively narrow conceptual realm of environmental management and public participation, onto broader questions surrounding the historical, political and cultural embeddedness of efforts to conserve nature. I also explore the relationship between the positionality and organization of territorial government, on the one hand, and the institutional character of policies aimed at managing natural resources and amenities, on the other. This discussion is situated within wider considerations of the political and social construction of nature during socialism and after. It is developed with reference to the emergent literature on the manner in which the inherent tension between path-dependent and path-creating policies is played out across space and place.

The chapter consists of two parts, the first of which commences with a survey of the main environmental legacies of socialist planning, consequently highlighting the mechanisms through which the economic and political logics of communist central planning affected the functioning of a range of different ecosystems. This is followed by a more specific discussion of nature protection policies, which – as was underlined in Chapter 2 – pre-dated the establishment of a Soviet-influenced, one-party system in CEE. I then review some of the main forces and developments that shaped the structures of municipal environmental governance in post-socialism.

The second part of the chapter focuses on developments after the fall of the Iron Curtain in the late 1980s. Having highlighted the important role of environmental NGOs in bringing about democratic change across many CEE states, I first focus on the radical transformation of state governance that these countries have experienced over the past 20 years. Particular attention is paid to the implications of local government reform, which has created a complex tapestry of spatial regulation across the region. The chapter then examines the manner in which environmental policies at different levels – and particularly at the local scale – have been implicated in the articulation of such processes. It is in this part of the chapter that I discuss the role of dynamics of territorial path-dependency, political fragmentation and economic neoliberalization on efforts aimed at protecting nature, and national parks. The conclusion of the chapter highlights some of the main gaps in the literature, especially within the context of questions addressed in the two case study chapters that follow.

The Environmental Legacies of Socialism

In order to understand the specific challenges associated with the governance of protected areas in CEE, it is necessary to highlight the broader path-dependencies inherited from the period of one-party rule, communist central planning and Soviet dominance. The environmental legacies of this era can largely be attributed to the economic development model pursued by CEE states during the same time: a system that has often been described as the 'administrative command economy'. Rather than market and price signalling, this apparatus relied on a series of government-controlled decision-making structures to plan and allocate the flow of resources. Not only did the state regulate all aspects of formal economic life, but it also was the sole owner of most assets, as a result of extensive nationalization programmes undertaken after the Second World War. The implementation of economic policies was also shaped and enforced by government bureaucracies, via targets, goals and plans set by the political leaders of government structures controlled by the Communist party. It is thus easy to see why and how the system stifled innovation, competition and investment in new technologies.

The rapid and extensive utilization of natural resources was one of the core tenets of such centrally planned economies. This is because the system prioritized economic production – effectuated via high rates of output, requiring intense levels of input – above everything else. In practice, this meant that the industrial extraction and processing of coal, oil, gas, uranium, hydropower, as well as ferrous and non-ferrous minerals were strongly supported and favoured by the state; the needs of heavy industry and the military-industrial complex gained precedence over the development of the service sector, or the provision of consumer goods. The availability of cheap labour (partly due to incarceration policies) meant that communist central planning was able to undertake extensive modifications of natural ecosystems, particularly in terms of watershed regulation so as to produce hydropower and regulate river flows. Such schemes were not limited to well-known examples in the Soviet Union. Even in Albania, for example, drainage and irrigation of the Myzeqe plain are considered by Rugg (1994) as one of the 'most impressive accomplishments of the communist regime' (60) since they reflect 'improved relationships between people and environment'. To support this argument, he underlines that 'between 1946 and 1990, 52,200 hectares of land were reclaimed from lakes, marshes, and swamps' while 'a further 200,500 hectares were improved' (ibid.). In total, the territory affected by the measures 'amounted to 36 per cent of the total arable land in Albania in 1990' (Rugg 1994: 60).

At the same time, the spatial and temporal mismatch between patterns of production and consumption created a specific economy of shortage, exemplified by the absence or scarcity of many basic goods, from meat products to automobiles. It should be pointed out, however, that the market was not entirely absent from centrally planned economies during the period of communist rule. Yugoslavia was notable for maintaining a system that was much more open to economic competition, private ownership, and bottom-up governance; a situation that can be attributed to its existence outside the Soviet political and economic sphere of interest. In the case of Hungary, Kornai (1992) established that approximately one-third of economic activity took place outside the formally regulated sectors. Overall, policies aimed at increasing the role of consumer demand and market relations in the mobilization and distribution of economic resources started to be introduced across the region – in one form or another – already from the 1970s.

Nevertheless, the heavy emphasis on economic production led to a situation whereby CEE states became extremely resource intensive in global terms. In 1990, for example, post-socialist economies needed up to four times more energy to produce a unit of GDP, compared to an average OECD country. In the energy sector, this situation was supported by the notion that the provision of electricity, for example, was a basic right guaranteed by the state, rather than a service associated with a unit price. An extensive system of cross subsidies was put in place to ensure that energy prices for households were kept artificially low. Not only therefore was the production of energy from domestic hydrocarbons strongly encouraged, but few provisions were put in place to address the environmental implications of their production chains.

It thus comes as little surprise that problems of air, water and soil pollution were pervasive across CEE, with far reaching consequences for the functioning of ecosystems and human health. Some of the most prominent issues included the immense scale of land degradation and harmful emissions – resulting, *inter alia*, in widespread acid rain that damaged vast tracts of woodland – associated with the extraction and burning of lignite in the notorious 'Black Triangle' at the borders of Poland, the Czech Republic and Germany. As pointed out by Bowman and Hunter (1991), perhaps the only environmental concern of similar transboundary proportions is the fallout from the Chernobyl accident, although the consequences of water pollution of rivers in the Danube, Elbe and Vistula watersheds also attracted significant attention. While the largest problems – in terms of spatial extent and environmental intensity – were concentrated in the territory of the Former Soviet Union, CEE states were also affected:

> For example, sixty-five per cent of Poland's rivers are unfit even for industrial use; eighty per cent of Prague's annual output of 40,000 tons of hazardous waste cannot be traced; forty per cent of Bulgaria's birds are endangered; and seventy-three per cent of the forests in the Czech and Slovak Federal Republic (CSFR) are severely damaged from acid rain. (Bowman and Hunter 1991: 923)

It is also important to note that the socialist era also left behind a number of processes and conditions that helped improve the environmental performance of CEE states. Most notable was the extensive investment in public transport systems, particularly in large cities; this was aided by the compact spatial structure of urban areas, whose development often followed a tightly-controlled pattern, due to the absence of private capital and a viable real estate market. Also of relevance in this context was the widespread presence of district heating networks. As emphasized by Ürge-Vorsatz et al. (2006), concentrated land use planning was one of the main reasons why towns and cities in this part of the world 'were easily serviceable by public transport' (2284). Basically, a dense settlement structure involving clusters of high-rise buildings 'provided ideal ground and economic rationale not only for public transport networks, but also for district heating networks' (ibid.).

Despite providing a potentially highly sustainable method to heat collective residential buildings, however, district heating has been in decline across many post-socialist countries further to the West (particularly in south-eastern Europe, where such networks have often suffered from vicious circles of disconnection). The problems faced by district heating networks illustrate the inherited technological issues faced by this type of infrastructure, as well as the inability of municipal governments – which often own the networks – to provide for effective means of public participation and economic management (Poputoaia and Bouzarovski 2010).

As I argued in Chapter 1, CEE states are characterized by high levels of forest cover and green space in cities; these features are a direct legacy of developments during communism, and can be attributed to a combination of several trends, including: i) the abandonment of rural and agricultural areas as a result of farm collectivization and urban industrialization; ii) the 'fencing-off' of large tracts of land for military and security purposes; iii) state-led afforestation programmes; and iv) eutrophication dynamics due, in part, to the inadequate application of fertilizers. Taff et al. (2010) have found 'unprecedented rates of farmland abandonment' that were accompanied by a process of out-migration

> … from rural areas to cities, and thus a substantial decrease in human impacts on the region's rural landscapes … this movement from rural to urban places commonly led to natural forest succession processes on abandoned farmland. (125)

The nature of political governance within the centrally planned economy, however, meant that public participation in environmental decision making was practically non-existent. Not only were citizens outside the party *nomenklatura* not consulted and/or involved in decisions related to the siting and operation of economic activities with adverse impacts on the environment, but the absence of formal civil society meant that the state displayed a complete lack of institutional capacity for democratic involvement in environmental policy decisions. These conditions led to a gradual groundswell of dissent over natural resource use and ecological pollution that would culminate in the late 1980s.

Nature Protection During Socialism

Before discussing the specificities of the nature protection regime in CEE during socialism, it is worth emphasizing that policies aimed at conserving particular areas and species existed in this region well before the start of socialism. As pointed out by Oszlányi et al. (2004), Central European countries possess a long nature protection and conservation tradition, dating back to the late seventeenth century. They argued that the roots of this process lie in the public realization that 'mining and industrial development creates severe environmental problems such as water and air pollution, deforestation and avalanches' (127).

Having emphasized that the world's first forestry university started to function in Central Europe, Oszlányi et al. (2004) argue that the notion of ecological sustainability has its origins in this part of the world, where it was 'applied in the sustainable production of wood in forests about 200 years ago' (ibid.). Indeed – and as was discussed in the previous chapter – various nature reserves were already established throughout CEE in the nineteenth century, often for narrow and utilitarian purposes; this process intensified after the First World War, when the number and scope of protected areas saw a rapid increase and fell under the patronage of the central state. The Soviet Union in particular established an extensive nature conservation system, aimed at curtailing human activities and protecting natural resources. The Czech case is indicative of such trends:

> A limited number of protected areas (mostly in the form of nature reserves) in Bohemia were established privately from 1838 onwards and a formal network of protected areas were created in the First Republic (including by 1938 c. 150 protected areas) under the auspices of the Ministry of Education and National Culture. (Tickle 2000: 65)

Some authors have argued that 'ecologically sound practices' in forestry and other kinds of ecosystems have existed in the region from as early as the eleventh century. Citing two examples from Slovakia – both dating back from the late nineteenth century – Oszlányi et al. (2004) emphasize that landowners 'also officially declared part of their land as excluded from harvest' being motivated by a desire to promote the 'maintenance and protection of the most precious parts of their land holdings' (ibid. 127).

However, the emergence of national parks *per se* can be traced back to a set of political and cultural traditions rooted in the romantic nationalism experienced by many European countries in the nineteenth century. This period was marked by efforts to establish and cultivate distinctive notions of national identity across CEE. The rising urban bourgeoisie of east Europe's capitals-to-be widely believed that the building blocks of a new conception of national belonging would be found in 'a primordial connection to nature' (Schwartz 2006: 46), which embodied deterministic thinking about the embeddedness of human culture and behaviour in the physical features of the natural environment. As a result,

> Writers and painters drew inspiration from folk themes and sought to depict the lives of peasants, who, thanks to their isolation in remote villages, were seen as having preserved the "pure" national culture and character, unadulterated by cosmopolitan influences, thereby sparing the nation from annihilation under the foreign yoke. (ibid. 30)

Therefore, CEE's national parks are inextricably linked to the concept of a 'nation of peasants', which 'has been a dominant trope since the national revivals of the nineteenth century, themselves fuelled to a large extent by the romantic fascination with folk culture and peasant authenticity. When they were formally established in the post-war period, many of eastern Europe's national parks still operated with some of these imaginaries – via a process that was either grudgingly tolerated or subtly accepted by communist ideologues – in order to justify their own creation. The designation of the first Soviet parks as 'national' rather than 'people's' is emblematic of this situation (Schwartz 2006). Other than Yugoslavia, no other Central European country used the 'national' prefix as a descriptor for these areas.

The complex ideological and political status of national parks in CEE meant that they were not uniformly present or governed across the region; while in some countries (such as Romania) this category of nature protection was practically non-existent, in others (particularly the Balkans) it was almost entirely dominated by the forestry sector. In the latter case, national parks represented little more than extended forestry organizations, whose principal role was to enact scientific practices of woodland management, while promoting tourism and constraining the activities of local people. In Romania, for example, Iojă et al. (2010) found that conservation planning mainly consisted of 'paper parks' before 1990. In a number of cases, national parks were established alongside tense international boundaries, where nature protection goals were secondary to military objectives. The common feature of all such areas, however, is that their development and governance were firmly controlled by the central state.

Territorial Government During Communist Central Planning

The legacies of the communist centrally planned economy in the environmental governance domain are also closely connected specific regulatory arrangements that existed at the local scale. The creation of a one-party system after World War II led to the complete transformation of the state administrative system in CEE. Sub-national government structures were completely reorganized to match the Soviet ideology and political approach. At the same time, 'territorial governments were established on the local level (rural and urban municipalities), district level, and regional (provincial) level' (Illner 1999: 10), with an organizational structure that involved 'an elected assembly, an executive board elected by the assembly and headed by a chairman, several committees composed of deputies, and an administrative apparatus'. Central government was given supremacy over basically all matters of regulatory importance, with sub-national units being relegated largely to utility and service management functions. Based on the available literature, it is possible to identify the following key features of local government under communism:

> The inexistence of accountability and democracy – despite the holding of formal elections at sub-national levels, these were mainly done for public relations purposes rather than fostering any meaningful democratic processes;

> The maintenance of effective power over all appointments and staffing policies by the apparatus of the Communist party, which meant that all officials in the line of management were vetted by its governing structures and there was a firm vertical line of command among all levels of administration;

> The insufficient economic autonomy of territorial government, which besides lacking its own property had its budgets completely determined by the central state;

> The inexistence of cross-sectoral co-operation and integration among the various branches of local government, and the corresponding central government departments. The vertical line of responsibility meant that there were overlaps in many domains, while others were insufficiently covered by state liabilities;

> The gradual emergence of local clientelism, as a result of the growing power of economic organizations as well as the expansion of relations of networking, negotiation and informal economic exchange. This meant that certain actors were more privileged than others, and effective action at the local scale was determined by the ability of individuals and firms to win favours from the political elites linked to enterprises and the Communist party;

> The divergence of national level conditions post-1960, as a result of reforms undertaken by individual countries (early 1960s in Czechoslovakia, early 1970s in Poland and Yugoslavia, and the early 1980s in Hungary).

Illner (1999) claims that a number of specific legacies arose out of this process. The policies adopted under communist central planning, according to him, created 'a separation of the private and the public spheres' accompanied by 'a popular distrust of institutions, of any political representation, and of formal procedures' (14). They also resulted in an 'unwillingness on the part of citizens to get involved in public matters and to hold public office'. At the same time, there was 'a paternalism that was characterized by a belief that local needs should be and will be taken care of by extra-local actors, usually by higher standing authorities' accompanied by 'a popular feeling of being chronically disadvantaged, of the community being neglected by authorities' (Illner 1999: 14). Overall, it can be concluded, local communities in most CEE states were in a position where a culture of pro-active local public participation was lacking at the end of communism; even though a burgeoning environmental NGO movement started to appear already in the late 1980s, citizens found it difficult to define the common interest on matters of local community concern, and to be proactive in lobbying the authorities for change.

Environmental Transitions in CEE

The late 1980s and early 1990s saw dramatic societal and political changes across the entire post-socialist space. Communist party governments were toppled throughout the region – often through dramatic revolutions – and replaced by multi-party parliamentary democratic systems and market-based economies. The movement towards a liberalized and market-based economy also led to massive structural changes, which also affected the state and regulation of environmental relations. Companies started to face 'hard budget constraints' as they could no longer be protected by the state via indirect and direct subsidies, and were forced to compete on open markets. Such developments led to the closure or downsizing of large industrial enterprises, particularly in the extractive and primary sectors, where most of the region's air and water pollution originated. At the same time, firms were forced to invest in more efficient and modern technology so as to catch up with their Western counterparts – a process widely assisted by the rapid entry of foreign direct investment into the region.

The aspiration to join the EU and meet international environmental standards provided a powerful impetus for the institution of new regulation and policy. It was widely believed that 'the pending restructuring of the Eastern economies posed an historic opportunity to find a new path toward sustainable development' (Bowman and Hunter 1991: 924). One of the earliest steps, therefore, was the introduction of air pollution measures to control sulphur dioxide, nitrogen dioxide and particulate emissions from large stationary sources. This led to a significant decline of sulphur dioxide emissions, sometimes surpassing 90 per cent of initial levels (Pavlínek and Pickles 2004). At the same time, declines of below 50 per cent in sulphur dioxide emissions were observed in the countries of the former Yugoslavia and Romania, while remaining countries in the region 'for which data is available recorded declines in excess of 50 per cent' (ibid. 244). Of no less importance was the drop in nitrogen dioxide emissions 'across the entire region, albeit less rapidly than declines in sulphur dioxide emissions' (ibid.). Where official data is available – particularly in Central Europe – there is evidence to suggest that particulate matter emissions also decreased, particularly in the 1990s. In less developed regions and countries, however, the decrease has been less pronounced, partly as a result of the growth in emissions from the transport and household sectors.

CEE states also saw a significant improvement in surface and groundwater quality thanks to declining levels of pollution. Not only did technological investment and industrial contraction decrease the amount of contaminated effluent discharged into the environment, but many states saw the construction of new wastewater treatment facilities, and the reduced use of industrial fertilisers and pesticides. These improvements, however, were uneven – the Balkans, in particular, still has significant problems with water pollution due to the lack of sewerage facilities, and the continued use of harmful chemical substances in agriculture. This region has only recently started to build effective systems for the purification of effluent water. In Central Europe and the Baltics, such policies were instituted already in the early 1990s:

> In Poland, the total amount of untreated industrial and household sewage discharged into rivers decreased by 70 per cent between 1990 and 1998, while the share of discharged sewage without any treatment in all discharged wastewater declined from 32 per cent to 17 per cent (GUS 1999). Only one per cent of effluent water discharged by industry was not treated in 1999 compared to 4.6 per cent in 1990. (Pavlínek and Pickles 2004: 250)

The post-communist transformation process was also accompanied by a significant decrease in the overall resource intensity of CEE economies, in relation to economic output. This trend can be largely attributed to the contraction of industrial activity – where over-employment and poor production practices were one of the main contributors to the distorted ratio of inputs to outputs – in addition to the movement towards serviced-based activities, the introduction of more efficient technologies, and the greater frugality of consumers and producers alike as a result of the introduction of cost-based pricing. Of particular importance was the rapid decrease in the energy intensity of the economy, thanks to the trends outlined above, as well as the implementation of state-supported programmes for investment in thermal insulation, the introduction of energy efficiency standards, as well as – in some cases – the development of labelling, tax, subsidy and incentive schemes. This was especially pronounced in Central European and Baltic states, where some countries managed to reduce their energy intensities by more than 40 per cent in Purchasing Power Parity GDP terms during the 1990s. Still, the fact that Russia remains the dominant source of hydrocarbons for many of countries in Central Europe means that they have often been forced:

> … to continue to rely on polluting, indigenous coal for much of their electricity supply, in the name of energy security. And yet these problems fade in comparison with the Balkan countries, which continue to maintain an energy-inefficient industrial structure. (Bouzarovski 2009: 460)

While the resource intensity of CEE states decreased in relation to economic output, overall rates of consumption have skyrocketed during the past 20 years, due to dynamics of economic liberalization, and the improvement of living standards. This is perhaps best illustrated by the rapid increase in automobile use – a trend that started already during the 1980s as a result of conscious attempts by several governments to increase the provision of consumer goods. At the same time, there was a significant decline in public transport use, and a shift from rail to road freight use. The most significant changes occurred in the decade between 1990 and 2001, when car ownership per capita practically doubled across most countries in the region, while public transport decreased up to 50 per cent in the most extreme cases. However, there is evidence to suggest that such trends have been gradually levelling out in more recent years (Pucher and Buehler 2005).

The relaxation of land use and planning controls also led to the expansion of suburban and exurban sprawl. The process has been supported by a range of factors, including: i) the establishment of a real estate market, thanks to property restitution and privatization policies; ii) the entry of foreign capital, which often preferred investment in out-of-town developments; iii) cultural preferences towards single family homes, automobility and shopping in large shopping centres, mainly as a reaction to the lack of access to such amenities during communism; and, iv) institutional dynamics – in the form of tax incentives, indirect or direct subsidies, or strategic planning documents – that supported the construction of new commercial, industrial or residential structures in greenfield locations. It resulted in the construction of numerous retail and housing developments at the outskirts of urban areas, often in regions that lack the adequate infrastructure to support them, or are ecologically vulnerable. The process has often resulted in the loss of valuable agricultural land, or the fragmentation of forest ecosystems; as will be discussed in the following two chapters, protected areas have also been strongly affected by such dynamics. The problem has received widespread media treatment in regions and countries with a longer experience of sprawl:

> As the post-communist transformation progressed, the environmental implications of suburbanization dynamics gradually started to gain the attention of the public and experts alike. It was pointed out that sprawl is present in areas where the quality of environmental amenities is high (Nuissl and Rink 2005) which itself suggested that suburbanization was leading to the deterioration of attractive physical environments at the urban outskirts. (Petrova et al. 2013: 1442)

In some cases, processes of liberalization and privatization, as well as the entrance of foreign direct investment, have worsened the environmental performance of post-communist economies. The relaxation of regulatory controls on multinational companies investing in the region has led to a number of large-scale environmental accidents and problems, including mining accidents resulting in severe water and soil pollution; and the continued operation of industrial facilities that release harmful emissions into the surrounding air, water and soil. Also 'privatization of forests and uncertainties over ownership rights and cutting regulations have led new private owners to cut their newly acquired forests for profit, ignoring environmental consequences and, in some cases, established legal codes' (Pavlínek and Pickles 2004: 256).

It should be pointed out that the last two decades have seen the flourishing of environmental non-governmental organizations – as advocacy, activist or community groups – across the region. The development of civil society in this domain commenced already during the late 1980s, in response to the widespread recognition that pollution and ecological degradation had reached critical levels (Sarre and Jehlička 2007). NGOs thus played a crucial role in the post-communist democratization process, often taking a major policy role at the national scale: some of the most prominent examples include the *Brontosaurus* movement in the former Czechoslovakia, and *Ekoglasnost* in Bulgaria. As the transformation process progressed, such groups became increasingly less grassroots-based, and more professional and goal-oriented. Some authors have been critical of such trends:

> The process of professionalization has isolated environmental interests from its grass roots, resulting in a loss of local knowledge and local input and making environmental interests politically vulnerable. (Baker and Jehlička 1998: 13)

However, despite being increasingly represented and conceived – in the media and public opinion alike – as 'non-representative and as part of the power structure' (ibid.), many environmental NGOs became closely integrated in the political process, either by being incorporated in party-led coalitions, or co-operating closely with official state structures. The latter development was also facilitated by the availability of international donor funding and the EU accession process. The Czech Republic's new state environmental administration, for example, promoted the role of particular actors 'explicitly by encouraging widespread consultation with NGOs over its policy proposals, funding new groups (partially through European Union PHARE monies)' in addition to 'ensuring consultative rights for civic and other interest groups in various elements of the new environmental legislation, including land use planning decisions within protected areas' (Tickle 2000: 70).

Nevertheless, 'despite the financial assistance, the access to transnational networks, and the additional opportunities and legal stipulations for environmental NGOs to become involved in deliberative policy and regulatory processes, the anticipated shift away from command and control and state- dominated hierarchic decision making has not occurred' (Carmin and Fagan 2010: 700). NGOs thus also played an important role in the integration of EU legislation in the national legal frameworks of CEE states that became members of this bloc (or are aspiring to do so). As a whole, this dynamic generated major changes in policy and practice, leading to a deep restructuring of the legal system and the institution of a wide range of standards and norms. It has also required the participation of numerous policy actors, with its associated challenges:

> The cases of Poland and Bulgaria demonstrate that nonselective local participation
> is not sufficient to ensure broad acceptance of international expert advice.
> Skewed representation of domestic political viewpoints in an international
> assessment can be as counterproductive as no domestic involvement at all.
> (Botcheva 2001: 221)

Having emphasized that, '[e]mbedding assessment processes in local institutions
that can provide for consultation of all important perspectives is critical for the
broad uptake of expertise' (ibid. 221), the same author singles out climate change
and biodiversity as 'global environmental problems' that are emblematic of the
participation of 'a large number of actors with different preferences' in this process
(Botcheva 2001: 221). It is thus easy to see why the approximation process was
associated with high compliance costs – in administrative, political and financial
terms alike – and required a fundamental reshaping of the institutional capacities,
relations and practices of relevant state organizations at a variety of scales. To a
certain extent, one of the main ameliorating factors was the fact that becoming
incorporated in EU requirements and procedures also increased the availability
of financial assistance and project support in the environment sector. As pointed
out by Carmin and Vandeveer (2004) the departure of many international donors
from the region has created a situation in which 'European intergovernmental
organizations and the EU have become the dominant sources of financial and
technical support for environmental policy change and remediation' (11).

This created a situation where most EU assistance to CEE countries in the
late 1990s prioritized the 'harmonization' of state policies and practices with EU
directives and regulations (ibid.). Prominent examples of relevant programmes
include the PHARE (Poland/Hungary Aid for the Reconstruction of the Economy)
Twinning programme – which focuses primarily on increasing capacities of public
organizations (Carmin and Vandeveer 2004) – and LIFE (Financial Instrument
for the Environment), involving the participation of Estonia, Hungary, Latvia,
Romania, the Slovak Republic and Slovenia (ibid.). However, not all authors have
appraised the EU's role in shaping the transformation process in positive terms:

> This influence has not been a democratic one and many contributors have
> pointed out that the involvement of the EU has displaced the centrality
> afforded to domestic concerns on the policy agenda, replacing it with those
> environmental problems most pressing for the EU, for example problems
> relating to transboundary air pollution. (Baker and Jehlička 1998: 24)

Such assessments correspond to earlier arguments that 'emerging democracies in
the region have taken a place in line behind all the other industrialized nations
in what is at best a slow trudge toward sustainable development' (Bowman and
Hunter 1991: 980). Nevertheless, there is widespread consensus in the literature
that the environmental conditions in CEE have been favourably influenced by the

EU's funding programmes, not the least thanks to the transfer of technological and management skills.

National Park Governance in Post-socialism

If there is one consistent theme that emerges from the limited literature on national park and protected area restructuring in CEE since 1990, it is precisely the lack of consistency. There simply is no consensus in the relevant literature – which, as was noted in Chapter 1, is limited in quantity, coverage and scope – regarding the extent to which there has been an improvement in the management of such areas. All of this is taking place under conditions influenced by neo-liberal transformation processes that are leading 'to uneven re-development accompanied by social and economic dislocation' (Tickle 2000: 75). At the same time, it has transpired that 'previous practices and networks' have been

> … influential in the production of new, post-socialist modes of environmental regulation, although changes engendered by western discourses and practices (principally emanating from the EU) are also strongly apparent. (Tickle 2000: 75)

Authors such as Pavlínek and Pickles (2004) generally see the situation in negative terms:

> When opened to the public after 1989, environmental quality was quickly compromised. In the 1990s, the management of existing natural [sic] parks was often negatively affected by efforts of private entrepreneurs supported by anti-environment-oriented politicians to reduce their size or remove restrictions on particular uses in order to allow economic activities such as logging and mining. (257)

These experts are particularly concerned about the environmental consequences of construction activities 'designed to profit from the rapid expansion of domestic and foreign tourism' (Pavlínek and Pickles 2004: 257) in protected areas. Their pessimism, however, is not mirrored by the group of authors who have seen the protected area implications of EU integration in positive terms. Work in this vein has emphasized the benefits of formulating and implementing a standardized set of criteria for the identification and creation of different protected area categories in a consistent and uniform way (Oszlányi et al. 2004). The financial benefits of EU membership have also been underlined in this context. Pullin et al. (2009) argue that 'agri-environment schemes can provide a significant source of funding by citing the example of Hungary; a country where €280 million were 'available for agri-environment schemes in 2006, which is approximately 15 times more than the state-allocated support of the Nature Conservation Authority' (820).

While arguing that 'large-scale policy drivers have the potential to change trends in biodiversity' (ibid.), such authors suggest that EU regulations, and the resources behind them, can be employed 'efficiently for nature conservation' and allow transition countries to 'retain much of their biodiversity' (Pullin et al. 2009: 820). According to Oszlányi et al. (2004: 133), this process has also been supported by the 'recently observed decrease in industrial air pollution' in addition to 'the introduction of low-pollution technologies, the improvement of energy efficiency in industry and housing as well as compliance with EU regulations' to generate a visible 'improvement in the ecological status of protected areas' (Oszlányi et al. 2004: 133). Nevertheless, the extent to which the legal and regulatory provisions stemming from EU accession can help lead to a long term resolution of some of the legacies from the past – as well as newly-created problems – remains unclear. Citing the 'interesting' situations created by conflicts surrounding the 'highway construction through the protected Biebrza marshes and Rospuda Valley and the illegal timber harvest in the Bialowieza area in Poland', Pullin et al. (2009: 820) 'believe that a focus on research support and policy pressure can have beneficial effects on biodiversity in these countries, despite the rapid economic growth'.

Some of the 'optimistic' authors emphasize that biodiversity loss can be addressed, in part, by providing the 'necessary financial and infrastructural support' to the 'body of well-trained experts with a wealth of natural-history knowledge' that exist within the region (Pullin et al. 2009). Much of the literature on national park and protected area conservation emphasizes the region's long tradition in 'the educated use of forests, ecologically sound practices in forestry and extensive agricultural production' in shaping post-communist practices of nature conservation (Oszlányi et al. 2004: 133). Lawrence (2008) has found that foresters in the region themselves act as major holders of expertise about environmental management, which is conferred not only by law and education but also 'gained through experience, and the acting out of an emotional commitment to the forest' (429). This suggests that:

> Because forestry is tied into histories of power and institutional culture as well as science and political rationalization, the evolution of forestry knowledge offers insights for wider debates about "expertise" as a socially constructed alternative to lay knowledge. (Lawrence 2008: 429)

A significant body of work in this thematic domain is focused on the Carpathian region, whose vast forest resources extend across the region, involving a range of protected areas – 'from small-scale nature reserves to large-acreage landscape protected areas, national parks and UNESCO biosphere reserves' (ibid.) – that cover between 9 and 32 per cent of the territories of all affected countries (Lawrence 2008). Even here, however, the last 20 years have produced markedly different outcomes; Kuemmerle et al. (2007) have found that rates of forest disturbance in Ukraine, Poland and Slovakia varied significantly during post-socialism, and were highly contingent on local institutional conditions. A similar situation has been found to exist in Albania, where the political and economic transformations brought about by postsocialism have 'altered rural resource values, changed the mechanisms through which forest users gain access to productive resources, and shifted the creation and distribution of resource rent among actors' (Stahl 2010: 140). By altering the incentives, decisions and practices of forest users, these developments have resulted in:

> ... an unprecedented rush on forest resources ... annual firewood extraction rates were three times higher than at the end of socialism. Consequently, forest users' practices became highly unsustainable and caused dramatic forest degradation. (Stahl 2010: 140–148)

As a whole, the evolution of protected area and national park governance during post-communism has brought into light the complex dynamics of power, knowledge and culture involved in the everyday articulation of nature conservation. This is confirmed by Schwartz's (2006) study of the changing narratives of nature protection in Latvia, through a case study of two Western-supported initiatives in national park management in the late 1990s, where she observes a conflict between those seeking to achieve 'national development only through an internationalist embrace of globalization and regional integration' (43), on the one hand, versus groups who see such 'narratives as threatening to agrarian notions of the value of rural landscapes' (Schwartz 2006: 43). She argues that the former discourse has 'appropriated the Western narratives of biodiversity and rural economic diversification' (ibid.), while the latter operates on the agrarian imaginaries of national identity that led to the original foundation of such areas. The role of representational strategies for political ends has also been highlighted in Franklin's (2002) study of practices of dispossession in Poland's Białowieża Forest, where biological scientists have impaired democratic processes via myths and imagery 'guided by romantic yearnings that hark back to schoolbook tales of royal hunting pageants that have, at their heart, primordial nationalist objectives' (1480).

Post-Communist Decentralization Processes: Challenges and Dilemmas

I have already highlighted, in several discussions within this chapter and the introduction, the complex changes in CEE's economic and political institutions brought about by post-communist restructuring. It should be pointed out, however, that this was not necessarily a rational process of constructing new structures in order to move toward optimal economic development goals. Stark (1996) contends that it involved '... rebuilding organizations and institutions not on the ruins but with the ruins of communism' (995). He argues that because of the importance of the 'ruins' of the state socialist system (in terms of trade, links, institutions, regulation, personal, and inter-firm networks), the economic transformation can be characterized as path-dependent. Still, this trajectory cannot be seen in deterministic terms, since actors in the transformation process operate within existing institutional constraints. As a result, some courses and horizons of actions were limited, while others were favoured.

The local government policies adopted by CEE countries during the transition were variegated and diverse. Overall, however, it can be concluded that there was a widespread process of political and administrative decentralization. The roots of this structural movement can be traced back as far as the 1980s, when local political action became a key focus for political resistance in countries like Poland. Despite the factors noted above – particularly the low overall development of a grassroots political culture – citizens generally wanted their governments to give greater power to decisions made at the local level. In part, some of these expectations harked back to pre-communist nostalgia, when the Austrian, German – and to a lesser extent, the Ottoman – empires that ruled this part of the world had developed extensive and functional systems of territorial management.

In terms of decentralization policies, the basic principle followed initially by most countries was to create small local government units that would reflect historical legacies and local peoples' wishes. Hungary was possibly the most successful in this regard, having established a functioning system already at the end of the 1990s. Other Central European countries were quick to follow. However, one of the main problems was the insufficient capacity at the middle tier of government – regional level governments often lacked resources, democracy, or both. Also present was the issue of political and economic fragmentation, due to the excessive number of municipalities in some countries. Work to resolve this is still in progress – while some countries still do not possess a middle tier of government with a properly functioning and clear mandate (especially in Central Europe). In others, there is still an effective lack of fiscal and political decentralization to the local level (particularly in the Balkans).

Such changes unfolded alongside the extensive transformation of post-communist cities and regions. Decentralization policies supplemented the breakdown of the centrally planned land allocation and urban planning model to intensify the processes of residential and commercial suburbanization and exurban sprawl that were discussed above. However, CEE municipalities have struggled to influence the management of public matters in this domain. In the Czech Republic, for example, the construction of thousands of satellite settlements – a specific form of exurbanization, with all the negative long-term impacts on the environment discussed above – happened at the behest of suburban municipalities who were eager to attract the population and tax revenue that accompanied this phenomenon. The poor capacity of regional and metropolitan government to regulate the matter, coupled with local authority fragmentation, meant that its larger scale consequences remained unchecked.

There was also an overall reorganization of the urban system in most countries, with national capitals or the centres of prosperous regions tending to expand at the expense of many small- and medium-sized towns in less prosperous regions. This process was catalysed by the uneven economic geographies of transition, which resulted in the concentration of high-value-added industries and services in a limited number of metropolitan agglomerations. 'Winner' urban centres (e.g. Prague, Budapest, Warsaw, Krakow, Ljubljana, Bratislava, Bucharest, Timisoara, Sofia) also saw dynamics of reurbanization and inner-city regeneration, while the effects of industrial decline and population outflow were most visible in 'loser' regions (such as eastern Poland, northern Bohemia and northern Moravia in the Czech Republic, the Great Plain and northern Hungary, eastern and southern Serbia, north-eastern Slovenia, northern Moldova and the south-eastern Romanian Plain, north central Bulgaria). The overwhelming majority of protected areas tend to be located in the latter, not the least due to being initially established in remote border and rural areas.

Nevertheless, it should be pointed out that municipalities have been taking an increasingly prominent role in managing some environmental policies in CEE. A central part of this has been the development of regional and local strategies for the sustainable development of regions and cities. As part of this process, for example, many municipal authority governments in CEE have developed programmes and strategies to address climate protection and energy security issues. These have been co-ordinated with wider spatial and urban development plans, as well as national building rules and standards. They have offered different options for the development of local infrastructure in particular. However, as evidenced by the implementation of Local Environmental Plans, the extent to which environmental policies are genuinely 'local' may be questioned:

> The available evidence suggests that such initiatives are part of a much wider network of social processes and interactions, the final outcome of which has been the devolution of environmental governance under the influence of international organizations and conventions. (Buzarovski 2001: 567)

In the case of Macedonia, it has been observed that such initiatives have been accompanied by a 'destatization' process 'orchestrated by the central government', which has 'proceeded steadily in spite of the problematic legal and economic situation of the local authorities' (Buzarovski 2001: 567). This can be linked to the broader difficulties –such as the lack of economic resources and effective political power – that have prevented municipalities from acting comprehensively in the domain of nature protection. What is more, the combination of rigid national policies and inadequate legal regulation has often created conditions in which local authorities are unable to affect infrastructural, technological and governance systems that fall under their territorial jurisdiction.

**Concluding Thoughts: Challenges at the Environmental Policy –
National Parks – Municipal Governance Nexus**

This chapter has explored the different ways in which nature protection practices and national park governance in CEE context have been shaped by the legacies of the communist past and the processes brought about by the post-socialist transformation, in terms of the regulation and implementation of environmental policy, as well as the capacity of relevant organizations to undertake effective measures in this domain. It is clear that the challenges associated by the post-communist restructuring process have deeply affected the 'institutional and economic parameters [that] determine the underlying vulnerability and adaptive capacity of societies' in the context of 'interventions and planned adaptations at the most appropriate scales' (Adger 2001: 921).

As a whole, the last 20 years have seen significant changes in the organization and governance of political relations and legislative systems in the environmental policy domain. Most countries in CEE – partly under the auspices of the EU integration – undertook a wide range of programmes to deal with the inherited communist legacies in this domain. Such policies led to significant changes in the management of environmental problems across the regions. Combined with the major decrease of resource intensity as result of broader developments in the economy, this has meant that much of Central Europe has seen noticeable environmental improvements in recent years.

The extent to which national park and protected area governance have improved, however, remains unclear. There is certainly sufficient evidence to suggest, however, that the external and internal pressures faced by protected areas have dramatically changed and increased during the last 20 years, often leading to the unravelling of conflicts that expose some of the tensions that lie at the heart of the initial establishment of such spaces. The character of these disputes emphasizes the importance of taking into account the social, cultural and economic practices that underlie the functioning of protected areas, particularly in terms of issues of democratic participation.

Given the diversity of the public sphere in CEE, it is difficult to make an overall assessment of its public participation abilities. What is clear, however, is that energy policy making in CEE is increasingly devolved to the local and regional level. Even though a new legal framework has been created to support this process, 'decentralization has not necessarily resulted in enhanced democracy or associated policy initiatives' (Tickle 2000: 77). Given the poor financial and regulatory power of regional government, as well as the problem of liability fragmentation at the local level, it is clear that a significant amount of capacity-building has been necessary in order to increase the administrative competence, institutional transparency and democratic proficiency of local authorities, especially in terms of their relationship with civil society organizations, enterprises and the public more generally.

The following chapters explore the relevance of some of these challenges to the operation of national park authorities in the context of wider environmental transitions underway in the post-socialist realm. In essence, I wish to move beyond the traditional remit of debates on protected areas – where much of the focus has been on nature-society linkages – onto an investigation of the ability of conservation organizations to process, as coined by Piermattei (2013), the 'plurality of perspectives and conceptions' contained within the given eco-political context. Also of relevance here are the mechanisms through which the legacies of past environmental governance practices (including those of the centrally planned economy, as the *differentia specifica* of CEE states) have influenced the ability of local populations to articulate effective communities of place and interest in relation to the process of nature protection.

Chapter 6

From Inflexible Legislation to Flexible Governance: Community Participation and Visitor-Related Challenges in Pelister National Park

Introduction

Not only have mountainous areas across the globe been shown to suffer from the same vulnerabilities that characterize their host societies, but it has also become evident that their unique geographic features make such areas sensitive to an entire additional set of social and environmental contingencies. This worldwide phenomenon also holds true in the case of Pelister National Park, which, in addition to being subjected to the broad level environmental legacies, trajectories and policies described in the previous chapter, has also found itself at the core of a range of specific circumstances stemming from the spatial and economic marginalization of the region.

It is against the background of the park's natural and social idiosyncrasies that this chapter sets out to explore the community participation and protected area management challenges faced by Pelister's local people, governing authorities and occasional visitors. I examine the micro-level articulations of the conservation practices that characterize the behaviours of the national park's local population and management body, with the aim of evaluating the 'real life' implementation of the co-management model for protected areas described in Chapter 3. Although various national-level institutions and consecutive governments have repeatedly declared their commitment to this approach as far as Pelister is concerned, there is limited evidence for its successful implementation on the ground. The chapter thus focuses on the multiple social, economic and political expressions of flexibility in national park management at the local scale, arguing that a 'bottom up' co-management model has already been *de facto* in place in Pelister for some time, despite the lack of a *de jure* framework at the level of the central state. I highlight the different components of this 'flexible' co-management model, while contrasting them with the inflexibilities that characterise Macedonia's formal legal system for national park management.

As discussed in chapters 1 and 5, Pelister is located in a part of the world that has experienced major economic, social and political change over the last 18 years. As such, it provides an unprecedented opportunity for studying the multiple interactions between rigid macro-level path-dependencies embedded in past and present policy documents and legal frameworks, on the one hand, and the flexible micro-level strategies and practices employed by local populations and management authorities as a method of overcoming such formal constraints, on the other. Of particular relevance in this context is the operation of regulatory frameworks for nature protection – especially with regard to protected area governance – and the manner in which environmental and economic policies have shaped the evolution of environmental management frameworks at the local scale.

Much of the evidence analysed in the chapter is drawn from on-site research, involving (as described in Chapter 1) in-depth interviews with local policy-makers, nature protection experts and local inhabitants, as well as questionnaire surveys of the residents of three villages in and around the park (Nizhepole, Malovishta and Brajchino). The data gathered in the field has been interrogated with the aid of a variety of qualitative and quantitative methods, and combined with a survey of the formal legal framework and policy documents for nature protection in the country to produce two analyses that highlight, respectively, the inflexible and flexible dimensions of national park management in Pelister.

These investigations, which form the core two sections of the text that follows, are preceded by a general overview of the human and physical geography of the national park itself. They are followed by an additional analysis that addresses the relationship between national park visitors and protected area governance, in response to the specific circumstances of Pelister: it is one of the most popular tourist destinations in Macedonia (alongside the country's other two national parks – Galichica and Mavrovo) while lying in the immediate proximity of several urban areas with a total population over 100,000. In the final section of the chapter, therefore, I explore the manner in which the motivations, perceptions and movements of national park visitors relate to the underlying dynamics of nature protection. Based on primary research undertaken in Pelister, I identify the profile of urban visitors to the park, while investigating their needs and aspirations. These analyses are compared with the results from similar case studies, and placed in the context of the broader relationship between nature conservation and tourism in the area.

Setting the Context: Salient Features of Pelister National Park

Pelister is the first national park in the Republic of Macedonia and one of the oldest such protected areas in the Balkans, having been founded as early as 1948. It is classed as an IUCN category II protected area. It is situated in the south-western part of the country, adjacent to the border with Greece and less than 15 km from Bitola, Macedonia's second largest city. Pelister encompasses the northern parts of the Baba mountain massif, extending between elevations of 891 and 2,601 metres above sea level (see Figure 6.1). Thanks to the expansion of its boundaries in 2007, the park now covers a total area of 14,300 ha. Although the area around the park is dotted by a number of villages – Brajchino, Kazhani, Rotino, Capari, Magarevo, Trnovo, Dihovo and Nizhepole – that are well known for their rich cultural and architectural heritage (see Figure 6.2), Pelister's territory only includes one rural settlement – the village of Malovishta – within its boundaries, since the park extends mainly across the upper parts of the mountain (see Figure 1.2). However, considering that the economies of all of these settlements are mainly based on tourism, agriculture and animal husbandry, their overall pattern of development and everyday life is heavily influenced by the path followed by the park.

Pelister's geological base is characterized by a unique combination of rocks from different eras, ranging from the Palaeozoic and Mesozoic all the way to the Quaternary. The heavily alkaline 'Pelister Granite', contained in a massive dome formation dating from the Ordovician, dominates most of the park and forms one of its key distinguishing features. This structure is embedded within a series of older, Palaeozoic green shales – another typical characteristic of Pelister's geological base. In addition, the park also contains Palaeozoic quartz- and quartz-sericite schists, as well as Mesozoic gabbro, dolerite veins, diabase and mermekitic granite. Glacial and periglacial landforms are among the main geomorphological attributes of the park, including a wide variety of relatively unusual – for this latitude at least – landforms such as cirques, moraines, granite block streams and fields, alongside nivation hollows, garlands, solifluction lobes, and ploughing blocks. Two of the cirques host tarns, which are well known throughout the country and represent a major tourist attraction.

Thanks to its varied geological systems, varied physical landscapes and pronounced mountain climate, Pelister has provided an optimal environment for the development of a wide variety of biotopes, including forests, dry grassland, mountain and freshwater ecosystems. As such, they comprise a diverse array of vegetation types, ranging from heath and scrubs to broad-leaved deciduous (oak and beech) and coniferous (Macedonian pine) forests. The park's numerous rivers, tarns and other aquatic habitats support a wide range of riparian communities, while areas above 2,000 m host alpine and sub-alpine grassland. While 9 out of Pelister's 32 different natural habitat types (9 forest and 16 grass communities) are protected by the Bern Convention as habitats that require special conservation measures – two of them are locally endemic communities. According to the Management Plan, the national park's key protection targets in this domain include:

> Species protected globally or in Europe: *Canis lupus, Felis silvestris, Lutra lutra, Myotis capaccinii* (mammals); *Rhinolophus blasii, Rhinolophus ferrumequinum, Rhinolophus hipposideros, Alauda arvensis, Alectoris graeca, Coturnix coturnix, Emberiza cia, Falco biarmicus, Lullulaarborea, Gypaetus barbatus, Perdix perdix, Pyrrhocorax pyrrhocorax* (birds); *Salmo pelagonicus* (fish); *Boletus regius* (mushroom);

> Species that are rare and protected in Macedonia: *Andreaea rupestris, Buxbaumia viridis* (mosses); *Crocus pelistericus, Gentiana lutea, Gentiana*

**Figure 6.1   Four granite peaks in the highest part of the
            once heavily glaciated Baba massif (visible
            here) reach heights of more than 2,400 m**

Note: Molika forests – sometimes interspersed with fir trees – grow
up to the treeline.
Source: Photo by author.

**Figure 6.2   The village of Brajchino is rich in
            architectural heritage**

Source: Photo by author.

punctata, Sempervium octopodes, Sempervium marmoreum, Knautia magnifica, Viola parvula (plants); *Achnanthidium kryophila, Luticola undulata, Navicula roteana, Pinnularia appendiculata* (algae); *Chroogomphus helveticus, Suillus sibiricus ssp. Helvetica* (mushrooms); *Parmelia exasperatula, Parmelia sorediata, Ramalina carpatica* (lichens);

Endemic species: *Alchemila peristerica, Dianthus myrtinerviu* (plants); *Niphargus pancici pancici, Eucypris kurtdiebeli* (animals); *Duvalius macedonicus, Duvalius peristericus, Tapinopterus comita, Nebria aetolica macedonica, Tapinopterus monastirensis monastirensis* and *Trechus goebli goebli* (insects).

Among the key distinguishing characteristics of Pelister National Park are the substantial Macedonian pine (*Pinus Peuce*) forests – locally known as 'molika' (Nastov and Micevski 1994, Nastov 2000). Covering a relatively large share of the park's northern slopes, Pelister's pine forests are among the Balkans' best-developed and most extensive ecosystems formed by this otherwise extremely rare and endemic pine (see Figure 6.1). The 'molika' forms two different types of vegetation communities on the territory of the park: mountainous woodland (*Digitali viridiflorae – Pinetum peuces*) found at altitudes ranging from 900 to 1,600 m above sea level; and sub-alpine woodland (*Gentiano luteae – Pinetum peuces*), usually present between 1,500 and 2,100 or more metres.

Layers of Institutional Rigidity: The Legal Framework for Governing National Parks in the Republic of Macedonia

The current direction of nature protection in Pelister is largely a function of broader developments in its host country – the Republic of Macedonia. Since gaining independence in 1991, and as a result of its aspirations to become a EU member, this country has continuously worked on the harmonization of its internal legal framework with EU legislation (European Commission 1996, 2005). Moreover, the universal right to a healthy environment is enshrined in the country's constitution, which obliges all of its citizens to respect environmental and nature protection principles. But these issues have traditionally received little policy attention and priority in the country, as evidenced by the fragmented and insufficiently comprehensive character of legislation relating to the management of protected areas, in addition to the lack of coherent, continuous, and integrated scientific work on the subject.

An important step towards the improvement of this situation was made in 2004, which saw the adoption of a new Law on Nature Protection. The Law, which aimed to achieve comprehensive natural heritage conservation, protection and management, allowed for the implementation of internationally recognized and ratified conventions in this domain, while transposing EU legislation regarding nature protection to the Macedonian context (GRM 2004b). In recent years, the legislative framework for nature protection has been expanded to comprise the Act on the Environment, as well as sectoral laws regulating the use of natural resources in particular domains, including fishing (GRM 1993), hunting (GRM 2004a), forestry (GRM 2013), pastures (GRM 2000), and plant health (GRM 2011).

At the same time, the state has taken numerous steps to ratify international conventions and agreements on nature protection. The long list of such documents includes: the UNESCO Convention concerning the protection of the world cultural and natural heritage; the Rio de Janeiro Convention on biological diversity; the Ramsar Convention on wetlands; the Bern Convention on the conservation of European wildlife and natural habitats; the Agreement on African Euroasian migratory wetland birds; the Bonn Convention on the Conservation of Migratory Species of Wild Animals; the Cartagena protocol on biosafety; the Agreement on Bat Protection (EUROBATS); the Convention on International Trade in Endangered Species of Wild Fauna and Flora (CITES); and the Convention for the Protection of Vertebrate Animals Use for Experimental and Other Scientific Purposes. Moreover, the Law on Nature has incorporated the two key European directives regarding nature protection: the Council Directive 79/409/EEC on the Conservation of wild birds and Directive 92/43 EEC on the Conservation of natural habitats of wild fauna and flora.

In addition to the increasingly thick file of legal acts, Macedonia has also formulated a formal Strategy and Action Plan on Biodiversity Protection. These documents are meant to provide policy instruments for expert support towards effective nature protection and management. While the Strategy defines the integral approach towards nature protection and sustainable use of specific biological resources, the Plan describes concrete activities that should be implemented in line with the main purposes of the Strategy. The Plan also contains a set of additional management tools, which should allow for the preparation and implementation of secondary regulation documents (i.e. protected area management plans), thus completing the legislative base for efficient nature protection and sustainable management.

One of the main purposes of the Law on Nature is to provide for the establishment and management of a system of protected areas aimed at maintaining biological and landscape diversity. Thus, the categorization and zoning of protected areas in the country has been implemented in line with IUCN (International Union for the Conservation of Nature) criteria. The Law defines six categories of protection: i) strict nature reserve, ii) national park, iii) natural monument, iv) nature park, v) protected landscape and vi) multipurpose area. It stipulates that each protected area may contain a zone of: i) strict protection, ii) active management, iii) sustainable use and iv) buffering. The total size of all areas protected through the Law has reached 187,770 ha (around 8 per cent) of the country's territory, including three national parks with a total surface area of around 108,338 ha.

As far as national parks are concerned, it is worth noting that the structure of their management system is highly hierarchical, with a two-way information system. Through the Ministry of Environment and Physical Planning (MOEPP), the government acts as an executive body responsible for the creation, management and protection of national parks. However, the day-to-day governance of such areas is carried out by national park authorities – public institutions established by the Government, in accordance with the provisions of the Law on Nature and the Act for the creation of the given national park. The Law on Nature also defines the structure of the authorities, which include a management board, director, an expert advisory board and a board for financial oversight. Their main responsibilities include: i) implementing the statutes of the national park; ii) adopting and implementing the national park management plan and annual programme and iii) formulating a financial plan for the park. The direct protection of the park is carried out by a specialized ranger service – established or designated by the park authority – while the efficiency of the authority itself, as well as the quality of the wider national park environment is monitored by the State Inspectorate for the Environment.

Therefore, it can be concluded that the three key factors that contribute to the 'background' inflexibility of national park management and protection are: i) the lack of protected area management plans; ii) insufficient human resources (as national park authorities are still set up as forest management companies in which environmental and social experts are absent); and iii) financial restrictions on their operations. The only flexibility of the current legislation on nature protection lies in the preparation and implementation of management plans for protected areas, defined and designed in accordance with the Law on Nature. The management plans, among other tasks, are supposed to provide opportunities for the efficient involvement of local communities in nature protection and management. However, the preparation of the management plans for protected areas is still in its infancy in Macedonia.

Nature Protection and Local People: A Diversity of Approaches

Macedonia's inflexible legislation and rigid legal management structure for nature protection has compelled the Pelister national park authority to develop and implement several alternative modes for effective local participation in the management and protection of the park. The authority's efforts in this domain have mainly been concentrated on the engagement of local stakeholders in park governance, while supporting them in the establishment of representative institutions relevant to the operation of the authority. Pelister's management institutions have also aided the creation of various organizational forms that may facilitate stakeholder co-operation and involvement in its activities. In addition to these 'top down' approaches, our research also indicated that the local population employs a number of 'bottom-up' flexible participation methods in order to improve its contribution to the decision-making process, while gaining economic and social benefits from the proximity of the park. In their entirety, both types of engagement have helped improve policy formulation and implementation in the park, partly by opening the space for the co-existence of different forms of knowledge and participatory frameworks related to protection of the national park.

The importance of the authority's role in the development of an effective management and protection framework for the national park – entailing the active involvement of local communities in its day-to-day governance – is all the more pertinent in light of the fact that policy practices in this domain are increasingly influenced by political and economic processes at larger scales (Hutton et al. 2000, Browning 2003). However, although global processes can be seen in locations where the livelihoods of people living in and around protected areas are influenced by global markets (Liu and Diamond 2005), local populations still play an important role in allowing protected areas to be managed in an effective and durable manner (McShane and Wells 2004).

Involving Local People via Top-down Approaches

The inflexibility of formal protected area management systems is mainly reflected in the limited use of natural resources, as well as the strict delineation of movement paths in the park. Such constraints may potentially lead to major conflicts of interest and land-use struggles between local people and the state institutions responsible for managing the national park. In this context, one of the key challenges for the national park authority has been the establishment of efficient top-down management structures that can satisfy the needs and interests of local communities, while providing for national park protection.

The results of the field research in Pelister pointed to some of the more successful approaches towards local community co-operation and involvement in national park management. In general, the evidence gathered though this process contradicted the widespread perception – common within part of the literature on the subject (Castro and Nielson 2004) – that a national park can only place a burden on the everyday life of local communities. Even though Pelister does possess a fairly rigid and externally-controlled formal management structure, it transpired that the national park authority has still found numerous flexible solutions, which allow local communities to benefit from its existence. Thus, the park provides a range of local services, including waste management, free fuel wood for heating, and the improvement and maintenance of local infrastructure – especially water supply systems – which all make a lasting and significant contribution to the strengthening of its relationship with the local population.

According to the Act on Nature, the national park authority is also responsible for the sustainable management and use of natural resources within the park. Considering that this includes non-timber forest products, it means that the Authority is fully responsible for organizing the trade of, *inter alia*, *Pinus peuce* seeds, cones and blackberries. In the case of the latter, the national park issues a specific number of licences to local families every year, allowing them to pick blackberries and sell them to licensed firms (see Figure 6.3). Although the fees

Figure 6.3   A Malovishta resident handles freshly-picked blackberries; a National Park vehicle and traditional village house can be seen in the background

Source: Photo by author.

for the licences are related to their types, several of the locals who we spoke with expressed concerns that their distribution may create local conflicts by favouring specific local villages and/or families, or engaging people who don't live in the national park area.

Another frequently used participation method has been the involvement of local residents as employees in national park management structures. As such, the practice has helped fortify the co-operation process between the local community and the park authority, especially in terms of improving fire precaution and protection, as well as preventing illegal forest logging. Thanks to this policy, a number of local inhabitants have been employed as full- or part-time members of the ranger service, or as foresters in the national park. In addition to such efforts, the park authority has also established and promoted local food and craftwork labels, beneficial for local communities in the development of alternative low-impact tourism in the park. Considering the significant rate of unemployment in the region (according to the Statistical Office of RM, the joblessness rate stands at around 35 per cent of the active workforce), it is hoped that development of nature-aware tourism in the park may increase the opportunities for investment in this area, preventing out-migration from local villages.

A key step in this direction was provided by the Project for the Conservation of Pelister Mountain, financially supported by the Swiss Agency for Development and Cooperation (SDC) in co-operation with MOEPP. This initiative, which was implemented in the Pelister area between 2000 and 2006, involved the establishment of special project offices in Skopje and Bitola under the support of the Swiss NGO *Pro Natura* – Friends of the Earth. One of its main purposes was the development of sustainable tourism, including rural tourism. During the realization of the work plan, it was concluded that Pelister's significant potential for nature-aware tourism is hampered by the park's poor infrastructure and the absence of skilled human resources. Wanting to improve the skills of the local population, the project team organized several different training and language courses in co-operation with the park management authority. They helped identify the need for official and efficient local community representation in the national park management, as well as the insufficient support for local NGOs as mediating institutions in this context.

Bottom-up Participatory Approaches

Unfortunately, top-down management approaches do not always ensure the significant direct involvement of local people in the governance of national parks (McShane and Wells 2004). Without feedback mechanisms, the authorities' efforts to foster a more meaningful co-operation process with local communities might result in an unsuccessful outcome (Stabler et al. 2009). Our field research in Pelister indicated that the local population has developed a number of bottom-up approaches towards the management of the park, allowing it to participate more actively in the Authority's work.

One of the most common of such models was the establishment of local NGOs, aimed at increasing the local population's role in the management of the park, as well as raising awareness about nature protection and the development of nature-aware tourism. These organizations have taken responsibility for a plethora of nature protection activities, including educational seminars, field trips to the forests, as well as preparing and disseminating informative materials. In order to achieve their aims, the NGOs also organise local festivities aimed at promoting traditional food, customs, cultural activities and craftwork. In Brajchino, families who are engaged in tourist services (accommodation, preparation of meals, guidance of tourists through the park) are organized in one NGO. They jointly define the prices for each service, thus reducing conflicts over price competition. Most interviewees emphasized that the NGO model has been very effective in terms of encouraging the direct participation of local people in national park structures, while ensuring that nature protection brings economic benefits for all of them.

> This organization has been functioning for a while now … and practically all of us participate in it. Actually there is only one villager who runs a B&B business without co-operating with us. Working together has helped us develop

a distinctive brand and become a foremost tourist destination in the country. ('Marija', B&B owner, personal communication, 16 August 2009)

In response to the opportunities for developing nature-aware tourism, many families from the villages in the area have rearranged their houses for tourist accommodation purposes, while establishing small enterprises for different tourist services. This strategy allows local populations to promote and develop rural tourism in a nature-aware manner, in addition to strengthening the local economy and empowering citizens to become a more influential factor in the management of the park. Furthermore, the direct dependence of their firms' revenues on tourism development – especially in terms of the number and quality of tourists – strengthens the local inhabitants' relationship with the national park. This is mainly because the growth of tourism in the area is directly related to the quality of its natural environment. Therefore, the close connection between nature protection and tourism growth provides a key incentive for local communities to participate in the projects and activities relating to the management and protection of the national park, led by relevant institutions.

The fact that nature-aware tourism is still developed unevenly and relatively poorly around the Pelister National Park area provides a useful starting point for any future analyses of the attitudes of local populations towards the park. The results of the research indicated that the scale of tourism development in the national park is not a decisive factor in the creation of local opinions and attitudes. All three villages involved in the research have developed local tourism at a different scale, depending on the number of tourists per year and the period of their stay (although the range of differences is valid only for local circumstances). In all three villages, the respondents expressed positive attitudes towards the national park's existence, mainly as a result of the improved investment opportunities that it brings.

Understanding the Socio-spatial Determinants of People-protected Area Relations: The Population of the Park

The favourable natural conditions concentrated in Pelister National Park, coupled with its geographical proximity to major population centres, have combined with the institutional thickness of local political interactions to facilitate the development of nature-based tourism and the trade in non-timber products in the area. In turn, such economic processes have started to influence migration trends; there is evidence to suggest that increased numbers of people have started to move to the villages in and around the park in order to exploit these opportunities.

The in-migration of new residents and capital into the region is most evident in the village of Nizhepole. During the past five years, this settlement has seen the construction of more than 40 new homes, often intended to serve as temporary holiday residences, or commercial purposes such as B&B accommodation (see Figure 6.4). Many of the houses have been built by pensioners who were originally born in the village, but who spent their working lives in western Europe. The village has thus become a focal point for the emergence of a specific remittance economy in the national park. One such returnee emphasized that:

> Having worked in the US for more than 30 years, I thought that Pelister would be a wonderful place to retire. I moved here with my husband three years ago, and we renovated the old family home. It is so peaceful here, and amenities are within close reach. ('Milka', pensioner, personal communication, 11 August 2009)

The area also attracted 'greenfield' investors from nearby towns who were quick to spot its tourism potential:

> I built a house a year ago … I want to open a B&B … I think that the location of the village is good for this. ('Jane', solicitor, personal communication, 7 August 2012)

These kinds of developments were supported by many locals from the 'old' part of Nizhepole, who saw ostensible benefits from the expansion of second home ownership and B&B accommodation:

> Thanks to the increased number of people in the village we finally have a shop and we do not have to go to the city to buy everyday essentials, including food. ('Kire', pensioner, personal communication, 14 August 2009)

Not everyone agreed, however. Several individuals who were born and lived in Nizhepole for their entire lives expressed concerns over the legal context in which the construction was taking place, as well as its environmental consequences:

> A significant number of new people are coming into the village … they are buying agricultural land and building large new houses without any building permission. These houses are outside the planning framework, there are no water and waste facilities for them … some of their owners throw rubbish into the river. ('Nada', shop assistant, personal communication, 9 August 2012)

Many people felt a strong attachment and sense of belonging in relation to the village and national park *per se*:

> We would not like to move anywhere else … we live next to such incredible and clean nature. I am not pleased with how the national park authority and local council function in the area, but the national park is here, and it is my home regardless of who is running it. ('Stojan', pensioner, personal communication, 14 August 2012)

**Figure 6.4   Nizhepole has recently seen the construction of B&Bs and
second homes at its outskirts**

Source: Photo by author.

While recent in-migrants are clearly behind the construction of the tourist
accommodation in Nizhepole, a very different situation can be encountered only a
few kilometres away – in Brajchino. Here, many local residents have demonstrated
an active interest in community and nature-based tourism. It seems that the NGO
mentioned in the previous section has played a particularly important role in this
context, by creating fair and equal tourism opportunities for everyone in the village.
Local people singled out the standardization of prices for accommodation, as well
as the availability of home-cooked meals and tourist guidance as the key benefits
from the existence of this organization. Many of them underlined that its existence
was made possible by the success of the Swiss-funded nature protection project:

> The whole idea started thanks to the "Pelister Mountain Protection Project" …
> What really makes me happy is that our NGO still exists and functions although
> the project finished two years ago. ('Elena', administrative worker, personal
> communication, 15 August 2009)

Aside from tourism purposes, Brajchino's residents also used the NGO for
improving their communication with the national park authority, and getting
involved in its management. This allowed them, for example, to become direct
participants in the decision-making process regarding the expansion of the park

into the hinterland of the villages, and the setting of new boundaries in the area. The close relationship between local public participation and place attachment and dependency in this context is evident in statements such as this by 'Panda', a pensioner:

> We are glad that the national park involved us in its policy process regarding the expansion of its border. After all, the national park is literally in our backyard and our lives and destinies are strongly connected to what happens there. (Personal communication, 14 August 2009)

The third village encompassed by this study – Malovishta – exhibits a radically different set of circumstances. Although it is located in a picturesque location with considerable natural and architectural heritage assets, this settlement's tourism infrastructure is at a nascent stage of development. The residents of Malovishta have been directly reliant on the park's natural resources and its management for their livelihoods, as the largest share of their income stems from employment in park services or the trade in non-timber products (see Figure 6.3). The flexible governance approaches adopted by the national park (mainly motivated by its desire to strengthen its relationship with the residents of the village) have further allowed this institution to support such activities via a range of policies, mainly thanks, as noted above, to the provision of a wide range of local services, and its willingness to tolerate and encourage a low level of subsistence-orientated activities focused on exploiting the forest resources of the park. Thus, unlike the remittance economy of Nizhepole, and the advanced co-operative model found in Brajchino, Malovishta still has a largely resource-based economy.

The local diversity of the economic and social circumstances of the three case study villages stand in contrast to the relative uniformity of their socio-demographic make-up. Overall, all three settlements are populated by individuals who are older, less educated, and more likely to be unemployed than the national average (see Figure 6.5). Their population has strong ancestral links to the area – 74 per cent of the surveyed individuals in the park stated that this was the main reason why they decided to live there. Also of importance in this context is the desire to live in the countryside (at 33 per cent), the inheritance of property (20 per cent), marital ties (13 per cent) and the nature of employment (12 per cent). While these figures explain the strong local embeddedness of the practices of Malovishta's inhabitants, they do not square easily with the entrepreneurial and co-operative spirit exhibited in Brajchino, or the high degree of capital influx and residential mobility seen in Nizhepole.

Perceptions of Environmental Conditions and National Park Management

Adding to the geographical differentiation of inhabitants in the park is a significant degree of diversity in opinions, perceptions and aspirations. For example, even

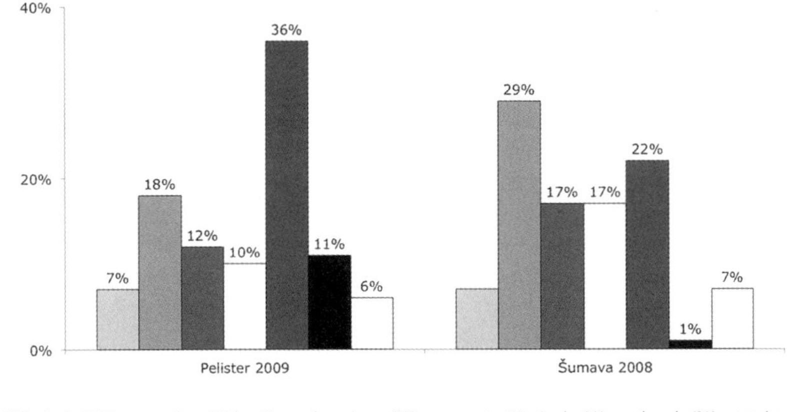

Figure 6.5   **Occupational structure of the survey samples in Pelister (N = 131) and Šumava (N = 183) National Parks**

Table 6.1   **Cross-tabulation of the variables 'village' and 'evaluation of the state of the environment 2006–2009' in different villages within Pelister National Park (N = 112)**

Place	Improved	Unchanged	Worsened	Don't know
Nizhepole	4%	54%	38%	4%
Brajchino	52%	39%	0%	7%
Malovishta	35%	65%	0%	0%

though 80 per cent of surveyed national park residents felt that 'the state of the environment has not changed' during the past 5 years, these opinions differed among three surveyed villages – with predominantly negative and positive views prevailing in, respectively, Nizhepole and Brajchino. Issues surrounding uncontrolled housing expansion in the former, versus the successful management of tourism in the latter, may have contributed to this picture (see Table 6.1).

The respondents were more divided when it came to the evaluation of current circumstances. In this regard, 71 per cent thought that 'the state of the environment is good', while 14 per cent felt it was 'poor' and a further 15 per cent did not know or have an opinion on this issue. According to almost two thirds of the respondents, the most serious problems relating to the protection of the environment stems from automobile traffic. The traffic issue might stem from the lack of a properly developed and adequately functioning municipal infrastructure, despite rising visitor numbers in recent years: this was singled out as a major environmental problem by an additional 24 per cent of the surveyed residents. The disposal and management of solid waste also ranks high on the list of local issues, as 15 per cent of Pelister respondents identified it as the primary environmental problem in the park. Almost two thirds of respondents did not think that the park faced any major nature conservation problems.

Although I applied a chi-square test to the examination of the relationship between the evaluation of the state of the environment and all the socio-demographic variables that could be extracted from the questionnaire, a statistically significant value was shown to exist only in a limited number of cases. The level of education was the only variable with a significant statistical value ($p = 0.0266$) in the appraisal of the environment of Pelister, indicating that respondents with a lower education were more negative in this regard.

A more nuanced view of the local population's understandings and experiences of environmental management and nature protection in Pelister can also be obtained via an exploration of the different ways in which the surveyed inhabitants' perceptions of the multiple dimensions of national park governance relates to their residential attachment to the area in which they live. In this context, it should be emphasized that around 90 per cent of the respondents from Pelister perceived the area as 'home', with a further 68 per cent seeing the initial designation and the current existence of the park in a favourable light. Also of significance in this context is the new Management Plan, which was drafted by the national park authority in 2004–2005 as the first document of its kind in the country. Even though the preparation and implementation of the plan might have contributed to the improvement of nature protection in the park, only 10 per cent of survey respondents had been included in its preparation, and 68 per cent of them were not even aware of its existence.

It is worth noting that while significant numbers of people were satisfied by the work of the National Park Authority as a nature protection organization and a management administration – the relevant percentages stood at, respectively, 43 and 36 per cent of all respondents – while 39 per cent felt the same when it came to its role as a cultural and educational organization. Yet 48 per cent of the interviewed residents thought that they could not evaluate the work of Pelister's governing institution in this regard, possibly due to the lack of interest or information. The statistical analysis of the respondents' answers to the different questions regarding the two national park authorities' work in correlation matrixes showed that there is a consistency in their answers. There was no relationship between the evaluation of the park's governing institutions and different demographic variables included in the survey.

Further analyses were undertaken by cross-tabulating the answers to the question 'do you feel at home in the national park' against those relating to: i) the functioning of the national park as a management administration, nature protection organization, and educational and cultural institution; and ii) the character of nature protection regimes. As was discussed in Chapter 2, my decision to select these indicators was based on the relevant literature, in which residential attachment to a given area can be considered an important indicator about the local residents' attitudes and behaviour towards a protected area. A number of authors working in this line of research have argued that populations with a stronger attachment to the place which is perceived as 'home' are more likely to be critical and interested about the current developments in it (see Kaltenborn 1998, Manzo and Perkins 2006).

It transpired that the respondents distinguished among their perception of the park as a home, designated protected area and management institution. Not only did this vary significantly according to the locals' place of residence (see Figure 6.6) but there were broader trends across the entire sample as well. Basically, the local residents' perception of the park as their home was not generally influenced by the level of (dis)satisfaction with the work of the national park authorities from the three defined aspects as management administrations (6 per cent of the respondents in Pelister were dissatisfied in this respect), nature protection organizations (where 8 per cent of Pelister negatively evaluated this aspect of the Authority's work), and educational and cultural institutions (the number of negative responses to this question reached 26 per cent of all respondents in Pelister). Although Pelister's residents were strongly attached to the park as their home, it seemed that they are not well introduced with, or have any interest in, the Authority's overall work, as half of them could not evaluate it. Also, their evaluation of the park as a protected area was not correlated with their (dis)satisfaction of the national park authority's work. The only clear connection was found in the case of the relationship between the nature protection regime in the park, on the one hand, and the Authority's work as a management administration (p = 0.0119) and nature protection organization (p = 0.0045) on the other.

The local residents' attitudes towards the austerity of the regime of nature protection in the national parks also revealed a number of underlying trends. In Pelister, approximately one half (46 per cent) of the respondents assessed the regime as 'appropriate', while most of the remainder could not provide an evaluation. This could indicate an absence of interest and insufficient information, or might stem from the fact that most of the local residents are in fact part time inhabitants of the national park. Stronger opinions could be found in the issue of how the park should be managed in the future. Here, nearly half (46 per cent) of respondents stated that tourist access to most endangered parts of the park should be permitted, while approximately a third (34) thought that it should be without any limitations, and only 21 per cent said that it should be prohibited. It is also worth noting that 73 per cent of the local people thought that possible future access to these areas should be 'with a professional guide only'. The other options chosen by the respondents were 'time-limited access' and 'charging an entrance fee'. At the same time, approximately 80 per cent of the respondents would have no objections to a further increase in further tourist and visitor numbers.

The surveyed inhabitants of the park were divided in their opinions about the influence of the national park on their everyday lives, as half of them agreed that this was true, while the remainder did not. Two thirds of the Pelister residents

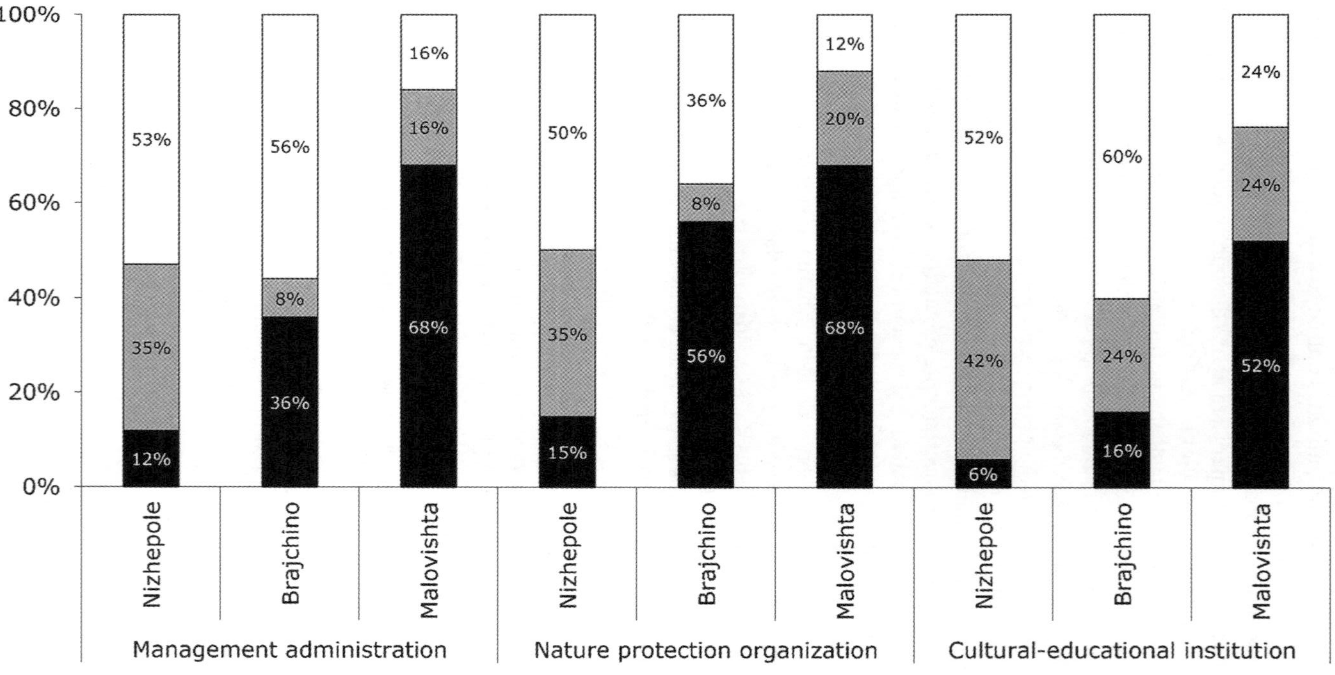

Figure 6.6   Respondents' evaluation of different aspects of the Pelister National Park Authority's work, according to their place of residence

Note: From left to right: N = 113, N = 112, N = 111.

who answered the question affirmatively felt that the park's role in their day-to-day existence could be felt in its ability to provide a clean environment, free fuel wood for heating during the winter, opportunities for tourism development and economic investment. But the lack of employment opportunities is clearly a major concern for local people; 60 per cent of respondents appraised this issue in a negative light. Respondents with a lower education and blue-collar employment were especially pessimistic (p = 0.0409 and p = 0.0215). At the same time, white-collar employees were more affirmative towards the claim that the establishment of the National Park has increased job opportunities in the region (p = 0.0441). As stated by a local interviewee:

> I have a US university degree – worked and lived there for many years before retiring and coming here. I think that being educated helped me appreciate and understand the benefits of a national park like this. It also made the area infinitely more attractive to live. ('Arta', pensioner, personal communication, 14 August 2012)

Most of the surveyed residents (85 per cent) stated that they had no economic profit from tourism in the park. Those who did enjoy direct or indirect economic benefits from this sector stated that they mainly did so via jobs that were dependent on, or related to, such services in different ways. Approximately 76 of the surveyed local people thought that tourist and visitor numbers had been increasing in recent years. Nevertheless, 86 per cent of the respondents stated that tourism is not increasing their day-to-day living costs.

There is evidence to suggest that the respondents' distinctive perceptions of the different roles of Pelister National Park are influenced by the specific historical and geographical position of the park. Most significantly, Pelister has traditionally never seen a strong clash of opposing interests with respect to the utilization and management of natural resources; as a result, the overall relationship between the local communities and the park authority in Pelister has not been riddled with a history of conflicts or antagonism. As pointed out by 'Petar', a village elder:

> You have to consider the fact that this area has never had a tradition of strong local government or local participation. We are located next to a major urban centre, which has held major political power and has been the focus of many important historical events. We have been its backyard, so to speak ... a place where there was a lot of population movement in the past, and a strong regional identity never took hold. (Personal communication, 17 August 2009)

Moreover, intense emigration processes have reduced the size of the local population to levels well below the carrying capacity of the park's ecosystems, which means that the surveyed communities don't really feel the pressure of Macedonia's strict nature protection legislation. This is further reinforced by the fact that the intensity of development in the area is still significantly lower than the limits prescribed by the national-level nature protection legislation.

Urban Visitors to Pelister National Park: Structures and Expectations

At present, tourism provides one of the key mechanisms for the articulation of Pelister's close relationship to the city of Bitola. As a protected area with easy access from the city, it serves as a 'sanctuary' for its urban dwellers. The broader literature on the subject – including work in psychology and natural science – argues that national parks, with their natural beauty and 'peacefulness' have made a significant contribution to urban residents' social and physical needs (see Suckall et al. 2009). In England, for example, the manner in which the countryside has shaped national identity over time has been extensively emphasized and discussed in relation to broader cultural, social and political developments (Agyeman and Spooner 1997, Edensor 2000).

The questionnaire survey in Pelister revealed that male visitors to the park outnumbered females by a ratio of almost 2:1, as their respective proportions in the survey sample reached 62 and 38 per cent. At the same time, most respondents belonged to the 40–59 age group (41 per cent), with a significant number being between 25 and 39 years of age (37 per cent). Those who were older than 60 or younger than 17 formed equal proportions of the sample (8 per cent each), with an even smaller share (6 per cent) being represented by the 18–24 age group.

Regarding the employment of the respondents, most of them (56 per cent) were working in the tertiary or quaternary sector, with a further 15 per cent belonging to the primary and secondary sector. It should be pointed out that 12 per cent of the respondents were unemployed, 11 per cent were students, and 6 per cent were retired. Nearly half of the respondents (45 per cent) had a tertiary education, while a further third (32 per cent) had managed to complete their secondary schooling. An overwhelming share (74 per cent) of the visitors were from Bitola.

When compared to the results of the European Agency for Reconstruction study (EAR 2003), it emerged that the only similarity in both cases was that male respondents outnumbered female respondents. Most visitors in the EAR sample were from the 17–24 age group, which also means that visitors with a lower educational level prevailed (52 per cent of all respondents had a secondary school diploma). It should be pointed out, once again, that this study was undertaken during the winter. As pointed out by 'Vesna', a 57-year-old private entrepreneur from Bitola:

> I don't come here in winter ... I don't ski and the hotels are usually full of noisy teenagers ... Me and my husband prefer Pelister in summer, the air is so clean and cool here. I cannot take time off from my business, only maybe on Sundays. Then we come here and have a coffee or lunch in the shade. (Personal communication, 20 August 2006)

Similarly, 'Vaska', a 45-year-old government employee, also from Bitola, pointed out:

> I would say that you get a particular demographic here in Pelister … mainly middle-aged people to pensioners. We are like a little community of regulars and its a big thing to be seen here, in the Molika hotel, drinking coffee, or taking walks in the area. It is good for your health. (Personal communication, 23 August 2006)

When asked about the frequency and length of their visits to, and stays in, the national park, 96 per cent of the surveyed tourists answered that they had already visited the area in the past. Although most of them pointed out that they usually visit the park during the summer (57 per cent), almost a third (34 per cent) stated that they tended to come to the park during the winter, with the remainder being split almost evenly between autumn and spring. Approximately 89 per cent of the respondents answered that they intended to stay only for one day in the park, while 7 per cent of them were staying for two days, and only 4 per cent were planning to spend a whole weekend in Pelister. It should be emphasized that 89 per cent of the respondents were using automobiles for transport to and from the park, while 8 per cent came by motorcycle and a mere 3 per cent by bike.

The main reason for almost all the surveyed tourists to visit the park were its natural features (98 per cent), although they did not demonstrate a special need for further exploration or educational activities. Most of them (76 per cent) pointed out that they would rather enjoy their meal or refreshment in the nearby restaurant in a natural surrounding. As indicated by 'Marko', a 32-year-old engineer from Bitola:

> Yes, sure, the moraines or pine forests sound great, but frankly, I don't have time to go up there … Golema Livada [authors' note – located at the entrance of the park] is so nice, you can drive here by car and bring your children to breathe some fresh air. Plus many of our friends come here too. (Personal communication, 24 August 2006)

It is interesting that the percentage of respondents who preferred 'peaceful relaxation' (90 per cent) was larger than the number (70 per cent) of those who came to the park for different sport activities. Out of all the interviewed visitors, only 3 per cent visited the park for cultural-educational activities. The sport activities identified by the respondents included hiking (57 per cent), picking forest fruits (26 per cent) and observing nature (17 per cent). Almost half (49 per cent) of the surveyed tourists thought that they had observed an improvement in the state of the environment of the park during the last five years, although 36 per cent felt that had stayed unchanged, while 4 per cent thought that it had become worse and 11 per cent didn't answer. Two thirds of the respondents were satisfied with the quality of the information system in the park, while a quarter were not.

Most of the respondents (87 per cent) felt that the most endangered parts of the park should be open to visitors without any restrictions.

The respondents underlined the proximity to the city (only around 15 km) and offered amenities, especially the well-known hotel-restaurant, as an important factor for the frequency of the visits. This came through in the interview with 'Bojan', a 67-year-old pensioner from Skopje:

> The nice thing about Pelister is that it is really close to the city of Bitola. It is practically a part of the city, an extension. But a nice extension, without pollution, without crowds. You don't have to walk very far to enjoy it. And you don't even have to have your own car … you can take a taxi and then walk up, it is not expensive. (Personal communication, 22 August 2006)

As a whole, therefore, the profile and expectations of urban visitors to Pelister corresponds with the common pattern of city dwellers visiting protected areas in Europe during both the summer and winter season, with some exceptions regarding the gender structures observed in Central Europe. It is important to note, however, that their presence has not, until now at least, exerted a significant impact on either the governance or natural environment of the national park. This is mainly due to the 'honeypot' effect created by the presence of attractive locations near the north-east entry points of the park, away from some of the area's more fragile natural locations.

Concluding Remarks

This chapter explored social and political relations arising at the local residents–nature protection–national park nexus in the Pelister National Park. Starting from the premise that local articulations of protected area governance in this part of the world are still insufficiently researched (despite work by authors such as Hall 2000, and Staddon 2009), I set out to scrutinize the political and institutional orderings that frame the activities of state authorities at different scales. I discovered that organizations in charge of national parks operate within the constraints of a rigid organizational and policy framework, subject to hierarchical decisions and tightly-defined competences.

Despite the inflexible legislative and management structure for nature protection, however, I identified several successful top-down and bottom-up approaches towards local community co-operation and involvement in national park management. These pertain to the availability of free services and investment in infrastructure, as well as the provision of full- and part-time employment in the national park. The national park has also involved local people in the trade of non-timber forest products, while offering various training and language courses.

As far as bottom-up approaches are concerned, it emerged that the area contains a number of participatory modes that allow local people to be actively involved in the National Park Authority's work. The most effective strategies of this type included the establishment of local NGOs as a tool for direct participation of local people in national park structures, the raising of public awareness about nature protection, and fundraising activities for local development. The creation of small-scale enterprises for the promotion and development of nature aware tourism was pointed out as a key factor for attracting 'high quality' tourists, thanks to its strengthening of the relationship between local people and the national park, and bringing financial benefits to both sides. The economic stability of the Pelister region is especially important in terms of preventing out-migration from the area.

The existence of multiple forms of 'adaptive governance' (see Chapter 2) is supplemented by a range of liberal policies in the everyday management of the park; this includes the lack of a strict delineation and regulation of movement paths in the park, which means that locals and tourists are free to roam throughout the area. The imperceptible presence of the national park in the day-to-day activities of local people – as opposed to its relatively clear institutional status – may have contributed to the locals' separate understandings of the park as 'home' and a 'protected area', on the one hand, versus a nature protection regime managed by a specific institution on the other.

Also of relevance to the relationship between local people and national park protection is the specific demographic structure of the area's residents; it transpired that Pelister is dominated by a socially homogeneous, relatively elderly and poorly educated population. This is linked to the fact that most of park's residents are either retired or homemakers. The majority of local people have lived in the area for a long time, and do not wish to move anywhere else. Adding to the strong sense of place attachment is the widespread appreciation of the amenities offered by living in the countryside, and the existence of powerful ancestral links (up to three quarters of the individuals in my survey sample possessed such connections). Aside from being visibly present in local customs and traditions, ancestral links also play an important role in the evaluation of local park management processes. Jobs related to the tourism sector are few and far between, owing to limited visitor numbers. But the low level of tourist development also means that pressure on nature protection and living costs from such activities is insignificant. At the same time, Pelister's inhabitants are, for the most part, positively inclined towards the increase of tourist and visitor flows in the park, despite having the impression that this has already been happening in recent years.

At the same time, the main motivation for urban dwellers' visits to Pelister also lies in the natural amenities offered by the national park, as well as the cultural brand associated with its image as a protected area. It should be pointed out that although the respondents in my visitor survey had a general knowledge about Pelister being designated as a national park, they lacked a significant awareness about their own impacts on the environment, and displayed a lack of readiness to take part in its protection. In part, this is also supported by the fact that the overwhelming majority of the visitors came by automobiles and only stayed in the area for a day.

The findings of the field work that I undertook in Pelister also support the suggestion that there is a strong relationship between the place where people park their automobiles and the facilities offered at that location. The main local road in the park actually connects Bitola with the largest and the most popular hotel-restaurant ('Molika'), which is also one of the main focal points in the area, and can be considered a gateway to the park. The implication of this finding for the management of visitor flows and national park protection is in line with Beunen et al. (2008), who have found that the creation of a gateway in a nature conservation area will help concentrate the traffic flows in the vicinity of the gateway.

Taking into consideration Eagles and McCool's (2002) argument that visitors from different age and social groups have different needs, one would have expected that most of the urban dwellers involved in my research would opt for physically easier and relaxing activities. The results of the survey confirmed this pattern, as 90 per cent of the respondents stated that they preferred 'peaceful relaxation' in the park. It is important to note that approximately two thirds of the surveyed visitors did not demonstrate a further interest in exploring or observing the natural features of the park. Clearly, the promotion of protected area tourism as a 'healthy lifestyle' option in Macedonia is making national parks fashionable for an increasing number of urban dwellers, without bringing about an associated rise in nature protection awareness. This emphasises the need for the development of an effective visitor management strategy as an integral part of overall national park governance.

Chapter 7
Šumava: Rewilding and its Discontents

I now move to the second case study, which focuses on the political conflicts and local participation challenges associated with the governance of Šumava National Park in the Czech Republic. As was noted in chapters 1 and 5, this protected area has seen rapid changes during the past two and a half decades, as a result of the dismantling of communist party rule and the movement away from the centrally planned economy. The opening up of the area to wider inflows of people and capital has increased its public visibility and status in the Czech polity. The park, which was already a part of the national imagination as an area of pristine natural beauty, has gained new importance as a foremost tourist destination. Yet the new level of public interest and economic investment has also created a series of tensions with local people over the content and nature of environmental management policies in the park. In addition, the raised profile of nature protection activities has also meant that issues of forest conservation have become a matter of national and international, rather than solely local or regional, concern.

The other important aspect of this park's current status stems from its specific geographic position: Šumava lies adjacent to international borders which themselves back on to protected areas in neighbouring countries. In particular, the south-western parts of the park are contiguous with the Bavarian Forest National Park – Germany's oldest protected area of its kind. This means that Šumava is invariably affected by developments across the border, and beyond the jurisdiction of national or municipal authorities. Such developments include both physical processes and nature management approaches. Superimposed on the park's complex environmental, economic and social geographies, therefore, are an additional set of challenges linked to its specific position with respect to the overflow of material and social processes occurring in other territories and spatial scales.

These two sets of issues – the tensions created by the changing political-economic status of the national park, and its permeability to wider processes beyond the boundaries of the nation state – form the primary object of discussion in this chapter. With respect to the former, I am especially interested in the manner in which local people have perceived, been affected by, and interacted with the actions of nature protection authorities. Of relevance in this context is also the functioning of other state bodies, the private sector and non-governmental organizations, all of whom have been shown to mould the development of protected areas such as Šumava. In terms of the latter aspects (the relationship between the national park and broader events), the chapter focuses on the political controversies generated by a bark beetle (*Ips typographus*) infestation of the park's spruce forests. As was mentioned in Chapter 3, some of the proposed measures to prevent the spread of the bark beetle across the park have generated heated public and scientific debates, thus exposing deeper dissonances between different belief systems and understandings of the relationship between nature and humans.

Given the embeddedness of the bark beetle controversy in wider nature protection issues, one of the main threads present throughout the chapter is the notion of 'rewilding' – which has been defined in Chapter 2. In Šumava, this change is visible in efforts to rebrand the National Park from the 'The Green Roof of Europe' into 'The Wild Heart of Europe' (Krenova and Kiener 2012). In the text that follows, I argue that the bark beetle has not only become the battleground for competing visions regarding the park's future development, but it is also acting as an agent for rewilding policies themselves.

It should be noted that such dynamics echo broader developments across Europe – a continent, which, until recently, was seen as a densely populated and industrialized space without any prospects for the creation of large protected areas with minimal human intervention. Understanding the driving forces and character of attempts to move towards wilderness-based conservation in Šumava can therefore provide important insights into the reasons for this shift in attitudes, as well as its broader political implications. The extent to which the specific historical and geographical features of the park – as well as its broader spatial location in post-socialist Central Europe – have made the region viable for rewilding by some policy actors, are also of significance.

Empirically, the chapter is based on analogous strands of evidence to those presented in Chapter 6: interviews with policy makers and local people, as well as a statistical survey with residents of the park. As a result of the prominence of national park management issues in Czech public debates, there is also a significant reliance on secondary data from news sources and policy reports. Before presenting the results of these analyses, however, I first provide an outline of the main natural and social features of the park, as well as the influence of past and present political developments on its environmental governance patterns. This is followed by an exploration of the broader institutional challenges encountered by the Czech Republic in the administration of protected areas. The empirical 'core' of the chapter consists of an examination of the changing demographic and economic structures of the park's local population, as well as its attitudes towards current nature conservation and governance practices. I then move onto an exploration of the broader political and ecological implications of public discourses surrounding the bark beetle's role in efforts to 'rewild' the National Park.

Salient Features of Šumava National Park

Šumava is the largest national park in the Czech Republic, with a total area of 69,030 ha. Just like Pelister, Šumava is a IUCN Category II protected area. The park encompasses a large part of the Šumava mountain region, at elevations ranging between 600 to 1,378 metres above sea level (see Figure 7.1). The park is surrounded by the much larger Šumava protected landscape area, which serves as an outer buffer zone and encompasses a further 99.624 ha (see Figure 1.3). Forest land represents 80 per cent of the park's territory, with only 0.1 per cent of its territory being dedicated to built-up areas. Šumava's geomorphological base mostly consists of rocks from the Primary and Palaeozoic age, making the area one of the oldest massifs in Europe. Remnants of glacial activity can be observed throughout the park, and particularly the higher peaks in its central sections.

It can be argued that the Šumava Region is one of the birthplaces of nature protection in Central Europe. The beginnings can be traced back to the nineteenth century, when in 1858, the Boubín Forest nature reserve was designated for scientific purposes in the Šumava Region. While further efforts to establish a national park in this region were undertaken in 1911, the main formal decision was undertaken in 1963, with the creation of a 163,000 ha protected landscape area focusing on the Šumava mountains. This development, which created the country's largest protected area at the time, was followed by the establishment of a 'Water Natural Accumulation Protected Area' in 1978, and the formation of a UNESCO Biosphere Reserve in 1990. The present-day park was officially created in 1991 by a special decree (No. 163/1991 Coll) adopted by the Czech Government. This area forms a continuous whole with the Bavarian Forest (*Bayerischer Wald*) Biosphere Reserve in Germany. It has been argued that the two reserves create the most extensive intact forest in Central Europe (Vacek and Podrázský 2003). Indeed, the combined territories of Šumava and the Bavarian Forest National Park have been certified as a Transboundary Park by the Europarc Federation.

The region's natural and social features have allowed it to receive wider international recognition, as evidenced by its inclusion in the European Natura 2000 network in 2004. The Šumava Peat Bogs (see Figure 7.1) have been declared a Ramsar site, while the area's geomorphological features – primarily glacial relics – are listed in the IUCN Red Book of ecosystems (Plesník and Roudná 2000). Other physical features of note include the dense network of streams and rivers, whose upper reaches are relatively unpolluted and allow for the existence of the common otter (Lutra Lutra).

The Šumava region is part of the Central European mountain forest biome, with dominant Norway spruce (*Picea abies*) and European beech (*Fagus sylvatica*) forests. Although in reduced abundance, the region also harbours communities of Silver fir (*Abies alba*), intermixed with Sycamore (*Acer pseudoplatanus*) and Wych elm (*Ulmus glabra*) (Jelínek 1985). The prevalence of spruce stands throughout the park is due, in part, to afforestation policies pursued in the nineteenth century, when monocultures of this species were planted over large areas. This process was an expression of the broader drive towards the heavy exploitation of natural resources, and the production of timber in particular. It resulted in the replacement of mixed broadleaf forests in the submontane and montane zone with uniform spruce stands, and resulted in the reduction of the park's formerly extensive virgin forests to a few upland valleys. The spruce monoculture, however, has recently

**Figure 7.1   A peat bog near the village of Borová Lada hosts
                  a variety of wildlife**

Source: Photo by author.

seen a rapid decline due to its susceptibility to wind and disease. Kyrill – a 2007 storm that caused the uprooting of spruce stands across an area of approximately 15,000 km^2 – was one of the main events contributing to the destruction of these forests, both physically and by aiding the spread of the bark beetle. The vulnerability of spruce stands to adverse weather conditions and pest outbreaks is further exacerbated by the even age and structure of such forests (Eriksson et al. 2006).

In phytogeographical terms, the Šumava region is defined as a province of the Central European temperate floristic zone (Vacek and Podrázský 2003). As a result of the high proportion of rare and endangered vegetation, Šumava National Park includes several 'Important Plant Areas', such as the mires of Modrava (7,893 ha), the Kremelna river basin (1,236 ha), the alluvial floodplain of the Upper Vltava/ Moldau River (2,432 ha) and Plesne Lake (82 ha) (see Vacek and Podrázský 2003 and Národní Park Šumava 2010). The park is also said to contain 69 species that are specially protected in the Czech Republic. These include the Heart-leaved twayblade (*Listera cordata*), nutgrass (*Scheuchzeria palustris*), Great sundew (*Drosera anglica*), Bohemian Gentian (*Gentianella bohemica*), Pannonic or Brown Gentian (*Gentiana pannonica*), Hairy stonecrop (*Sedum villosum*), Yellow bog sedge (*Carex dioica*), Yellowishwhite bladderwort (*Utricularia ochroleuca*), Yellow pond-lilly (*Nuphar pumila*), Mud sedge (*Carex limosa*), Creeping sedge

(*Carex chorrdorrhiza*), Boreal bog sedge (*C. magellanica*), Slender cottongrass (*Eriophorum gracile*), Burnt orchid (*Orchis ustulata*), Southern adderstongue (*Ophioglossum vulgatum*), Early or Yellow coralroot (*Corallorhiza trifida*), Frog orchid (*Coeloglossum viride*), and Alpine cottongrass (*Trichophorum alpinum*) (Národní Park Šumava 2010).

The park also contains a number of important animal species, including the Eurasian lynx (*Lynx lynx*), European otter (*Lutra lutra*), Western capercaillie (*Tetrao urogallus*), Black grouse (*Lyrurus tetrix*), Hazel grouse (*Tetrastes bonasia*), White-backed woodpecker (*Dendrocopos leucotos*), Ural owl (*Strix uralensis*) and Corn crake (*Crex crex*). Other typical fauna includes Red deer (*Cervus elaphus*), Garden dormouse (*Eliomys quercinus*), Northern birch mouse (*Sicista betulina*), Northern bat (*Eptesicus nilssoni*), Alpine shrew (*Sorex alpinus*), Black stork (*Ciconia nigra*), Three-toed woodpecker (*Picoides tridactylus*), Ring ouzel (*Turdus torquatus*), Boreal owl (*Aegolius funereus*) and the Eurasian pigmy owl (*Glaucidium passerinum*) (ibid.). It is important to note that lynxes, which were extirpated by the mid nineteenth century, were re-introduced into the area during the 1980s, via an approved release of 17 lynxes into the wild between 1982–1987. Subsequent monitoring has shown the population exceeding 50 individuals, and migrating freely across the border to Germany. The two national parks (Bavarian Forest and Šumava) currently represent the core lynx habitat, and the centre from which the animals are spreading to new areas. The lynx is also the only natural predator of the park's thriving deer and elk populations, which are kept in check by occasional culls.

The wider region that hosts the national park is one of the most sparsely settled areas in the Czech Republic, with a population density of only 1.3 inhabitants per km^2. A total of 22 settlements are partially or fully located within or around the boundaries of the park, although six villages – Borová Lada, Kvilda, Horská Kvilda, Srní, Prašily and Modrava – lie in the immediate proximity of many of its tourist attractions (see Figure 1.2). The low population density of the area is mainly a result of the dramatic economic and social changes that it witnessed throughout its recent and more distant past. The area was associated with a resource-based economy in the past, capitalizing on the region's significant mineral and forest resources. Although the earliest signs of organized human settlements in the region can be traced back to the eleventh century, the intensification of economic activity and anthropogenic impacts did not commence until the seventeenth and eighteenth centuries. This period saw the expansion of animal husbandry, wood processing, mining (mainly iron and gold) as well as glass making in the region.

The twentieth century was marked by gradual outmigration trends towards nearby urban areas, mainly located in the prosperous regions of Bavaria and Bohemia. But World War II led to a radical shift in the region's development trajectories. Two waves of population displacement – involving Czechs at the beginning of the war, and the German-speaking majority at the end of war – led to the reduction of the number of residents in the area to a third of levels seen in 1900. The abandonment of many settlements and the destruction of local

**Figure 7.2   A fragment of the border fence that separated the former
                    Czechoslovakia with the Federal Republic of Germany during
                    the Cold War provides a potent reminder of the Iron Curtain**

Source: Photo by author.

economies were further exacerbated by the establishment of the Iron Curtain in
the immediate proximity of the park (see Figure 7.2). This resulted in the creation
of military training areas and border control zones that large parts of the region
almost inaccessible for 50 years (Tickle 2000).

Nevertheless, the natural features of the region – coupled with population
policies promoted by the central state – meant that a low intensity of in-migration
was achieved during the socialist period (see Furlong 2006). Recreational and
tourist uses were also present during this time; their intensification post-1990
led to a rapid inflow of capital and population into the park. In short, the last
two decades have once again completely reversed the economic fortunes of the
area, turning it into one of the Czech Republic's prime areas for real estate and
tourism investment. When placed in their historical context, these developments
mean that Šumava is currently faced with an entirely new set of demographic and
economic challenges.

Organizations in charge of running the park have exacted a significant influence
on the governance of current policy conundrums stemming from the legacies of
the past. A single authority is in charge of managing both the national park and
protected landscape area. It possesses a multilayered top-down structure with a
director appointed by the central Government. The authority has traditionally been

organized along the lines of a forest enterprise, as its activities have historically been concentrated on forest management and timber trade. In 2006, following the broader efforts of the Czech Republic to adapt its legal acts in line with relevant EU standards (Cihar et al. 2000, 2001, Cihar and Stankova 2006, Furlong 2006, Kušová et al. 2002, 2005, 2008, Petrova et al. 2011) the authority underwent a major process of organizational reform, which resulted in the establishment of six departments in place of the previous 10 forest administrations. This restructuring process was aimed at transforming the authority from a forest management company into an organization which, while taking responsibility for forest management, will create a basis for considering nature conservation and sustainable local development through an integrated framework. Public relations and marketing departments were added to the authority in order to improve its communication and co-operation with local stakeholders.

The park is formally divided into three zones, in line with the protection regimes in place: these range from 'core' areas where no human intervention is allowed, to zones surrounding human settlements and the borders of the park that both have seen, and are currently experiencing, considerable anthropogenic transformations. The fact that zones under strict protection do not form a contiguous whole means that the general model of development and everyday life in the park is profoundly affected by the policy decisions made by the park management authority. The importance of local concerns came to the fore in 1991, when the borders of the park were defined based on a compromise between expert assessments, on the one hand, and the aspirations and viewpoints of local people, on the other. Although the state is the main owner of the park's land (approximately 85 per cent; local municipalities own around 10 per cent and the rest of the land is private), some of the municipal forests – a valuable resource for local development – are spatially situated in the centre of the park. This has complicated the zoning and management of nature protection.

In order to contextualize some of the discussion that follows, it is worth noting that the park has experienced a continuous change of management practices throughout the period since its formal establishment. This has been accompanied by numerous political conflicts between the managing authority and the resident population of the park. As was noted before, decisions over the future development paths of the area have constantly been the subject of high profile political debates. Some of these discussions have stemmed from the lack of regulatory legitimacy associated with the designation of the park. Contrary to common international practice for Category II protected areas, the national park was originally established via a government decree rather than a custom-made legislative act (which would be common international practice for for Category II national parks in IUCN). The first legal act on Šumava Park was adopted by the Czech Parliament only in 1991, but not without its own share of controversy: populist politicians, local people, scientists and civic associations all hold widely differing views on the direction and content of particular policies with respect to the management of the park.

Nature Protection in the Czech Republic

Of key relevance to developments in Šumava is the wider state of nature protection activities in the Czech Republic – a country that, as was noted before, has transformed its policies in this domain to a significant extent during the past two decades. The beginnings of nature protection in the Czech Republic can be traced back to the first half of the nineteenth century. The country's first nature reserves – Žofin and Hojná Voda Forests – were designated in 1838 for romantic, aesthetic and ethical reasons. Scientific purposes were the main reasons behind the establishment of the first forest reserve in Šumava – Boubín – in 1858. The twentieth century saw a rapid intensification of such processes: the first Protected Landscape Area, a category equivalent to the international IUCN Category V was established in 1955, followed by the adoption of the first Act on nature protection in 1956. Krkonoše National Park (V IUCN category) became the first Czech national park, with its designation in 1963. The protection of plant and fauna species was significantly intensified during the 1970s and 1980s, leading to the publication of the first Czech Red Lists of threatened species (Plesník and Roudná 2000).

After the 'velvet divorce' experienced by Czechoslovakia in the 1990s, the Czech Republic posited EU accession as an overarching policy goal. This was followed by the harmonization of the Czech national legislation with that of the EU in all fields of society, including nature protection as well. For example, after becoming a European Union member, the Czech Republic managed to transpose the *acquis communitaire* into its national legislation. By adapting itself to the relevant legal sources of the EU, the Czech Republic has made the necessary efforts to be a full member of the Union as far as environmental and nature protection issues are concerned (e.g. Natura 2000). This policy effort has been followed by academic research relating to the nature–society relationship and/or tourism development in protected areas (Hall 2000, Cihar et al. 2001, Cihar and Stankova 2006). However, while all of this work contributes to the general knowledge about local residents, nature protection and national park management in the Czech Republic, what is usually missing is the comparison and interpolation of existing information with data from similar analyses in central and south-eastern Europe.

As a result, Act No. 114/1992 Sb. on the Protection of Nature and Landscape was adopted. Currently, this is the most important instrument regarding nature and landscape protection in the Czech Republic (Furlong 2006). The Act provides both general and special territorial species protection. Regarding the general species protection, this Act ensures a legal protection of all flora and fauna, including the protection of wild birds and species of trees growing outside forests. It protects them from activities, which might endanger their existence or cause their degeneration, disrupt the reproductive ability, and bring about the species' population extinction or the ecosystem destruction. General territorial protection pertains to the entire territory of the Czech Republic, including the defined territorial systems of ecological stability, important landscape features, character of landscape, natural parks, and provisionally protected areas.

At the same time, species that are under a special protection regime by the Act have been categorized into three groups, in accordance with the threat level associated with their extinction, ranging from 'critically endangered' to 'highly endangered' and just 'endangered'. Specially protected fauna and flora species are specified in Annex II (plants) and Annex III (animals) of a government decree (No. 395/1992 Coll). The Act also provides special territorial protection for a number of 'Specially Protected Areas' that are deemed unique in terms of particular geological, biodiversity, or cultural aspects. It specifies six categories of such areas, including national parks, protected landscape areas, national nature reserves, nature reserves, national nature monuments, and natural monuments.

It is also worth noting that the Bird and Habitat directives (79/409/EEC and 92/43/EEC respectively) were transposed into the Act on the Protection of Nature and the Landscape as a result of the Czech accession to the EU. Thus, two additional types of nature reserves were established in line with the EU Natura 2000 network of protected areas: Special Protection Areas and Sites of Community Importance. Overall, the Czech Republic currently possesses 29 large protected areas, including four national parks and 25 protected landscape areas. The estimated size of all protected areas, including those under the Natura 2000 framework, accounts for 18 per cent of the country's territory. Outside the four national parks – which are all associated with separate authorities – nature protection in the country is managed by the Agency for Nature Conservation and Landscape Protection of the Czech Republic. All of these organizations are legally and organizationally embedded within the jurisdiction of the Ministry of the Environment. This means that, similar to other countries in Central Europe, nature protection in the Czech Republic is an institutionally centralized and vertically integrated affair.

Local People in Šumava: Demographic Structure and Opinions About the Environment

The survey I undertook with Šumava residents (described in detail in Chapter 1) indicated the presence of a younger, more educated and less unemployed population, in comparison with Pelister (see, for example, Figure 6.5). National differences notwithstanding, this discrepancy points to the ascending economic status of Šumava in the Czech Republic and CEE more generally – a situation that lies in stark contrast with the recent history of the area. The survey also revealed that only 17 per cent of the respondents in Šumava were retired, while more than a third worked in the tertiary sector; it can be argued that figure points to the existence of an employment structure associated with tourism-related services.

Approximately one half of the interviewees in the park had an ancestral connection to the region, demonstrating the presence of strong local ties and heritage (in contrast to some of the discourses described in Furlong 2006). This is further supported by the fact that one quarter of respondents cited marital ties as a key residential pull factor in favour of the park. Yet the inheritance of property played a minor role in shaping the residential choice of interviewed locals, as only 7 per cent of them answered this question affirmatively. Taken in the context of the high percentage of people with ancestral and marital ties, this figure suggests that most of my respondents were second-generation migrants. At the same time, nearly a third of the surveyed local people (32 per cent) stressed that the main reasons for their choice of the park as a place to live were associated with the nature of their job. The amenities offered by life in the countryside also played an important role in boosting Šumava's attractiveness, with 23 per cent of respondents identifying them as primary reasons for their continued residence in the area. Despite these contingencies, the evaluation of place attachment to the park appeared to be weaker when compared to Pelister (even though approximately 90 per cent of respondents still broadly identified with the park as 'home' – see Table 7.2).

It also transpired that the interviewed residents held a highly diverse set of views with respect to the state of the environment and nature protection in the park. This finding is in line with the controversies associated with the adoption of new regulatory documents in recent years, as well as the reported lack of satisfaction among local people with the ways in which state authorities were managing the park. Nevertheless, 40 per cent of the respondents in Šumava thought that the overall state of the environment had remained unchanged between 2003 and 2008, with 29 and 27 per cent saying that it had deteriorated or improved, respectively. In terms of the current state of the environment in the national park, the majority of respondents (54 per cent) provided a positive assessment, as opposed to the 30 per cent who appraised it negatively; almost 10 per cent stated that they did not have an opinion in this regard.

Just like in Pelister, almost two thirds of the respondents in Šumava thought that automobile traffic was the main causal factor behind the current environmental problems of the park. The excessive intensity – partly as a result of tourism growth – of car and bicycle transport in Šumava might have contributed to the residents' opinions in this regard, especially in light of the fact that almost 20 per cent of my respondents identified cycling as the main environmental problem in the park (also see Figure 7.3). As pointed out by 'Karel', a teacher:

> There has been an explosive growth of car traffic in the area – and the good roads that take you to nearly all corners of the national parks do not help. There is a good bus network but the authorities should do more to invest in this. (Personal communication, 15 August 2013)

Issues around the management of solid waste were seen as one of the park's primary environmental problems by 15 per cent of surveyed locals. As far as nature conservation is concerned, almost half of the respondents thought that the main issue in this respect was the quality of the forests as a result of the bark beetle

Table 7.1　Answers to the question 'Do you feel at home in the park region?' (N = 175 for Šumava, N = 117 for Pelister)

National Park	Absolutely yes	Relatively yes	Relatively not'	Absolutely not	I don't know
Pelister	83%	12%	3%	2%	0%
Šumava	47%	43%	4%	2%	4%

Figure 7.3　Cyclists are a common sight on the roads of Šumava's villages, which otherwise contain a diverse mix of architectural styles

Source: Photo by author.

'calamity'. A cross-tabulation of variables describing the nature of employment and local people's evaluation of the state of the environment yielded a significant value (p = 0.0061), while retirees and blue-collar workers were shown to evaluate the state of environment in a more negative manner.

When asked about the regulation of tourist admittance to the 'strictly protected zone' of the Šumava National Park 60 per cent of our respondents stated that access should be forbidden, while the remaining 40 per cent thought the opposite. It is also worth noting that 54 per cent of the respondents thought that possible future access to the most endangered areas of the park should be 'with a professional guide only'. Yet these figures did not correspond to the local residents' views on the influence of the national park on everyday life and employment. While the share of surveyed residents who evaluated Šumava's influence in this regard in a positive manner reached only 35 per cent, the overwhelming majority of respondents (62 per cent) who provided a negative answer pointed to the limited freedom of movement and use of the natural resources, as well as strict building regulations. It should be pointed out that 40 per cent of the surveyed individuals assessed the regime of nature protection in Šumava as 'strict', with 30 and 9 per cent, respectively, thinking that practices in this domain were 'appropriate' and 'relaxed'.

The local residents' negative opinions on such topics may be attributed to the lack of employment opportunities in particular. Approximately 70 per cent of the respondents stated that there is a shortage of jobs in Šumava; only 16 per cent had the opposite opinion. Nevertheless, 34 per cent of the respondents thought that the establishment of the national park had increased the number of job opportunities, while 16 per cent did not agree with this statement. As pointed out by 'Jana', who was unemployed, it was the nature of employment that matters most in this context, rather than the number of jobs in itself.

> The work that is available for local people is mainly in low-pay and low-skill jobs. If you have a good education and are interested in starting up or working for jobs that require a higher degree of specialization, you are better off migrating to Prague or Western Europe. Even though we are a prime site for tourism and scientific research the best jobs in these domains do not go to local people. (Personal communication, 10 August 2013)

Working-age respondents with blue-collar jobs were more likely to appraise the existence of job opportunities in Šumava negatively (at $p = 0.0110$ and $p = 0.006$, respectively). It is interesting to note, however, that working-age individuals were more affirmative towards the claim that the establishment of the National Park has increased job opportunities in the Šumava region ($p = 0.0105$), as did white-collar employees in both parks (as evidenced by $p = 0.0087$). Approximately 40 per cent of respondents thought that tourist and visitor numbers had been increasing in recent years, with an overwhelming majority of the park's inhabitants (93 per cent) stating that tourism is leading to a rise in living costs. Still, 80 per cent of the surveyed locals in Šumava had no objections to the intensity of tourist flows on the footpaths in the park; the same percentage decreased to 62 per cent in the case of tourist movements in the vicinity of their homes. But only 2 per cent of them stated that they would have no objections to an additional increase in tourist and visitor numbers. There was fierce opposition to increasing traffic flows in particular:

> There are just too many tourists in the park. They are everywhere – you can't even get out of the house without running into them. I don't know who is worse: the cyclists or the motorists. While the motorists emit noise and pollution, at least they are restricted to the roads. The cyclists you see everywhere, even on paths that are not supposed to be used by them. ('Jacek', pensioner, personal communication, 14 August 2013)

Local People's Attitudes towards, and Relationships with, the National Park Authority

Almost all of the statements and opinions of local residents in Šumava pointed to the decisive role of state-led nature protection practices in determining the articulation of political and environmental issues in the park. To a certain extent, this can be attributed to the dramatic economic and social turnaround experienced by the park during the past two decades. Environmental management practices have moved from a highly interventionist approach – involving the cutting and removal of vulnerable tree stands – towards a 'natural', non-interventionist strategy aimed at fostering the self-regeneration of the forest. Both phases have excluded the local residents in different ways. In the first instance, forest exploitation was subcontracted to external companies and took place in areas that were outside the reach of the park's inhabitants, while the later phase, *inter alia*, has led to the reduction of employment opportunities for local residents as full or part-time woodcutters. At the same time, paths of movement through the park are strictly defined, and the right of entry into the first zone of 'strict protection' remains prohibited. As pointed out by Sofie, a B&B owner:

> We wouldn't mind it so much if the first zone was off-limits to everyone, but it is not fair that local people can't even walk there while people from the authority are allowed to drive guests in huge jeeps through them. Of course we hear the same excuse all the time, that all of their visits to the first zone are for scientific purposes. ('Sofie', B&B owner, personal communication, 7 August 2008)

It should be pointed out, however, that the authority has managed to implement a number of measures that are seen favourably by local residents. These include the maintenance of cross-country ski tracks, the establishment of 'ecological' buses, the construction of information centres and the rehabilitation of public landmarks in the park (Kušová et al. 2008). There have also been efforts to improve the state of waste and wastewater treatment facilities. Still, according to residents such as 'Stanislav' (a pensioner) the park authority is more interested in pleasing tourists rather than locals:

> They are doing things only for the tourists. We, the older, retired people are just left on our own. There is not even a pharmacy close enough, so I have to buy my medicines in the city, but for the tourists there are information centres in every village. ('Stanislav', pensioner, personal communication, 13 August 2008)

In line with the extensive body of literature on the value of the metaphor of 'home' as an indicator of deeper social and cultural affiliations (see chapters 2 and 6), a significant part of my work in Šumava – just like in Pelister – scrutinized the similarities and differences between the residents' perceptions of the national park as 'home', on the one hand, in relation to their reactions and attitudes towards the work of the national park authority, on the other. The questionnaire interrogated three aspects of the national park authorities' operations: as management administrations, nature protection organizations, and educational and cultural institutions. Almost half of the respondents (42 per cent) were satisfied by the National Park Authority's work as a nature protection organization, although only 27 per cent thought the same when it came to its management role. The work of the Authority as a cultural and educational organization was evaluated favourably by 64 per cent of the respondents. It also transpired that respondents with a higher education evaluated the work of the Authority as a management administration and nature protection organization more favourably (at $p = 0.492$ and $p = 0.0157$, respectively). Moreover, respondents with no ancestry from the region, as well as those whose families were living in the park for more than three generations had more positive attitudes towards the overall work of the Authority ($p < 0.001$). As pointed out by 'Klara':

> I was born here, and have lived in this village for more than 30 years. Historically we have had excellent relations with the national park. They provide jobs for many of the people living here, and have been responsive and willing to listen to many of our suggestions and concerns. I think that people who moved into the area more recently, to buy holiday homes, are more detached from local politics and decisions. ('Jan', forester, personal communication, 11 August 2008)

Comparing the residents' evaluations of Šumava as their home, on the one hand, and a national park authority with three different functions, on the other, indicated that this area was in a slightly different situation compared to Pelister: The residents, overall, had stronger opinions about the functioning of the park. This is despite the fact that the structure of answers to the questions about the three different roles of the park were mixed (ranging from 'very satisfied' to 'very dissatisfied' in almost equal proportions: see Figure 7.4). There were significant geographical differences among the villages in this regard, with residents in Horská Kvilda and Srní – villages with pronounced local dissatisfaction over tourism development and unemployment, respectively – evaluating the park's role as a 'management administration' in a generally less favourable light (see Table 7.2).

I also found a strong correlation between the perception of Šumava as 'home' (otherwise 90 per cent positive, as noted above) and one of the aspects of the National Park Authority's work: the level of satisfaction with its role as a management administration ($p = 0.0258$). This effectively means that respondents did not make a distinction in their perceptions of these two dimensions, and can be considered more critical than the ones from Pelister (thus demonstrating the validity of the theory about the correlation between place attachment and attitude creation). Moreover, the evaluations of the park's general existence in the area, as well as the quality of its protection regime, were significantly correlated with the overall level of (dis)satisfaction with the National Park Authority's activities as a

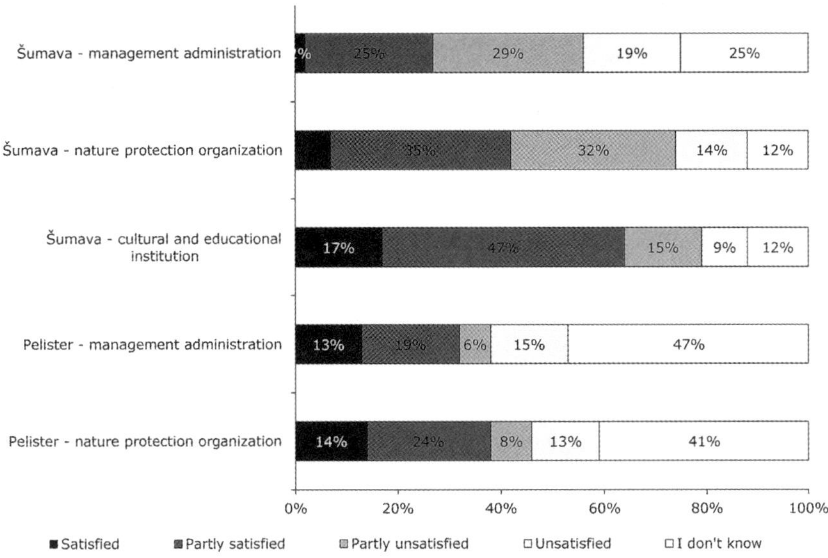

Figure 7.4　**Evaluation of Šumava National Park Authority's work as a management (N = 183), nature protection (N = 182) and a cultural-educational institution (N = 182); and Pelister National Park Authority's work in the same domains**

Note: N = 113, N = 120, N = 111, respectively.

Table 7.2　**Evaluation of Šumava National Park Authority's work as a management organization in relation to the respondents' place of residence (N = 182)**

	Satisfied	Unsatisfied	I don't know
Horská Kvilda	13%	60%	27%
Srní	17%	63%	38%
Prašily	23%	61%	16%
Kvilda	26%	44%	31%
Modrava	29%	65%	6%
Borová Lada	55%	31%	14%

whole (including all three aspects of its operations), since p < 0.001 in both cases. This means that local people who held strong viewpoints about the operation of the national park were also likely to be more opinionated with respect to its overall presence and policies.

The Bark Beetle: A Rewilding Agent

The aforementioned discursive shift in the representation of Šumava from Europe's 'green roof' to its 'wild heart' has deeper roots in socio-ecological processes currently underway in the park. It is linked, in the main, to the emergence of a specific policy of non-intervention during the past 10 years, largely in response to the wholesale destruction of spruce monocultures as a result of the expansion of the bark beetle (*Ips Typographus*). Debates over the extent to which this infestation should be managed by humans in order to preserve the forest resources of the park exposed the deeper political ideologies and allegiances of relevant stakeholders in the area, as well as the relationship between local people and the natural environment in the context of management practices associated with the park's governing bodies in the past.

The bark beetle is considered one the most threatening pests to mature spruce stands throughout Eurasia (Grodzki et al. 2004), being responsible for the destruction of up to 9 million m³ of timber per year (Schelhaas et al. 2003). Yet this species requires a complex combination of factors in order to expand in a given forest. The temperature of ambient air, for example, is said to regulate the bark beetle's swarming activity, diapause and winter mortality (Hlásny et al. 2011). The temperature regime during autumn can also play a crucial role in influencing the abundance of the swarming population during the following spring (Jönsson et al. 2009). It is argued that the bark beetle mostly attacks old or forest stands damaged by wind, as they provide substantial quantities of breeding material (Grodzki et al. 2004). Especially vulnerable are spruce stands located outside their natural habitats – mainly at higher or lower elevations. Heat and drought stress, in particular, can affect trees that grow on below-normal elevations (Jump et al. 2009).

The first documented bark beetle outbreak in Šumava's territory dates back to the nineteenth century. In 1870, a storm with hurricane-force winds devastated a significant part of the area, levelling numerous spruce trees. Woodlands affected by the hurricane provided a fertile breeding ground for the bark beetle, which soon expanded beyond their borders and destroyed both the remaining spruce monocultures and part of the surviving primeval forests. The twentieth century saw numerous outbreaks, the most prominent of which happened in 1996. It is argued that this epidemic originated in the neighbouring Bavarian Forest National Park, where a thunderstorm destroyed 175 ha of spruce forest almost a decade earlier, in 1983 (see Krenova and Kiener 2012). Having infected the fallen stands, the bark beetle spread throughout the park and across the border into Šumava. Despite such developments and protests by local people, however, the management of the Bavarian Forest National Park decided to allow the affected areas to develop spontaneously, rather than clearing them. This, in effect, led to the establishment of a new non-intervention management model, predicated upon the notion of rewilding spaces historically affected by human activity (Meyer et al. 2009).

At the time, there were roughly three non-intervention areas in Šumava (a number of infected trees were salvaged). They encapsulated a territory of approximately 2,300 ha, including Modravské slatě, Trojmezná and a small area around the Prášilské Lake (see Figure 7.5). In the rest of the core zone of the Park, each infested site was treated on a case-by-case basis. During the early 1990s, bark beetle cleansing was permitted in principle, but the National Park Authority could decide how to deal with the infected trees. In other words, the rewilding of Šumava was placed in the hands of the top management of the Park. It is therefore not surprising that the Park's policy and attitudes towards the bark beetle have fluctuated in accordance with the preference of each Park's director, and the closest collaborators associated with this post (which, as was noted above, is subject to a top-down appointment by the central Government).

In 1995, the then national park director Ivan Žlábek – who was not favourably inclined towards non-intervention policies – decided to re-arrange the zonation of the Park, even though he was in post for only one year at the time. Not only was the core area decreased by half as a result, but it was also divided into 135 individual plots. The Park's management justified this radical decision by arguing that the bark beetle will not be fought in all 135 sites. Yet infested trees were felled in 53 such areas in 1999. Moreover the alleged non-intervention regime did not ban the conduct of afforestation activities in the core areas. The Management Plan, approved in 2001, complicated the situation even further by dividing the core areas into additional sub-categories. The non-intervention regime was seen more as an exception rather than the default strategy undertaken by the Park's managing authority. The splitting of the core area into islands and sub-categories has created numerous conservation and management issues related to the rights of local people to use and move through the land, and the implementation of rewilding policies in 'core' areas.

The next crucial phase for the rewilding of Šumava was, to a certain extent, influenced by the Kyrill windstorm. As was noted above, this natural disaster uprooted thousands of spruce trees and spread the bark beetle infestation throughout the Park in 2007. Subsequent evaluations of particular ecological processes in the affected areas showed that the plots where bark beetle-infested trees had been actively logged in the recent past were extensively damaged (Krenova and Kiener 2012). This was taken to signify that management practices in the Park need to be changed. The post-Kyrill devastation was also seen as an opportunity to create a large Czech-German forest landscape, thus helping implement an early twentieth century plan for a transboundary protected area. The Czech Republic's accession to the EU (2004) and Schengen treaty (2007) would provide a further boost to this initiative, otherwise interrupted by the two world wars and the Iron Curtain. Thus, building on a 1991 memorandum for collaboration, the Authorities of the Šumava and Bavarian National Parks – as well as the ministries responsible for environmental issues from both countries – signed an agreement for the unified governance of almost 14,000 ha of core area in 2001. It was agreed that 64.7 and 45.3 per cent of this space would be located in, respectively, Germany and the

**Figure 7.5   Prašilské Lake is a major tourist spot in
             Šumava National Park**

Note: The non-intervention zone can be seen in the background of the photo; bark beetle
damage is evident at the top of the ridge above it.
Source: Photo by author.

Czech Republic (Martin et al. 2008). The consequent 'Wild Heart of Europe' was
created and defined as a:

> large area of high biological value and intactness, which is mostly undisturbed
> by man, roadless and without other industrial infrastructure or permanent
> habitation, and where extractive activities are not allowed. Visitor access is
> permitted on certain hiking trails and guided tours are preferred. Scientific
> research is permitted but no manipulative experiments are allowed. (Krenova
> and Kiener 2012: 121)

Management guidelines for the Wild Heart transboundary area were formulated
in 2009; the collaboration between the two parks also resulted in several joint
European Union-funded projects, such as Interreg, Leader, and the German-Czech
Future Fund. The resolve to establish a non-intervention regime in Šumava was
further strengthened by the entrance of the Green Party in the Czech Parliament
in 2006, as well as the appointment of a well-known proponent of rewilding
policies – František Krejčí – as the park's Director in 2007. The promotion of a

non-intervention policy in the Šumava was also supported by the Second European Congress of Conservation Biology that took place in Prague in 2009.

However, the Park does not only border German protected forest areas, but it also lies adjacent to Austria's mainly privately-owned and commercially-harvested spruce forests. The friendly treatment of the bark beetle in the German and Czech national parks was not seen in a favourable light on the Austrian side of the border. Thus, during both 2008 and 2009 the then Austrian Minister of Environment sent letters to his Czech counterparts with a request to stop the spread of the bark beetle to the Austrian forests. As a result, an agreement for the monitoring and control of the bark beetle's transboundary expansion was signed in 2009.

However, following the Green Party's loss of their parliamentary seats in 2010, the Ministry of Environment was taken over by the conservative Civic Democratic Party (ODS). Both the top management of the Park and its governance practices saw yet another rapid change. The new director – Jan Stráský – declared an all-out war on the bark beetle using all available means, including heavy machinery and chemicals. Clear cuts started to be undertaken even in areas that had previously been designated as non-intervention zones, and hunting was allowed in the entire core area. This policy saw little change following the appointment of Jiří Mánek as the park's director in July 2012. He continued an active battle against the bark beetle, using similar approaches to those of his predecessor. These were in line with the Ministry of Environment's new Act on Šumava National Park, adopted in 2012. It advocates the decrease of core areas and the promotion of a more relaxed policy towards economic development projects in the tourism domain (ski lifts and slopes in particular).

A common thread links all such events: the agency of the bark beetle has been persistently mobilized in discourses surrounding the socio-ecological future of the park. In a way, this organism has become an integral factor in the governance of the park's socio-ecological system. But perspectives on its origin, rights and impacts have been divergent across different public debates, as well as local government and political parties.

On the one hand, the bark beetle has been presented as an indigenous, local and natural inhabitant of Šumava's spruce forests, whose life course and expectancy do not only depend on the park's governance regime (whether it involves interventionist or non-interventionist) but local climatic conditions and the structure of the forest as well (Jonášová and Prach 2008). Natural scientists, conservationists and environmental NGOs have thus not seen anything problematic about the dynamic co-existence of the bark beetle with the park's forest ecosystems. Rather, they have argued that the beetle can lead to the transformation of such woodland from a spruce monoculture dominated by spruce stands into a dynamic, complex, mixed forest that will be more resilient not only to insect infestations but to changing weather conditions more generally (since strong and destructive windstorms are quite frequent in the Park). The bark beetle has also been depicted as a symbol of the aspirations of Czech science to demonstrate its ability to both keep up with and further develop contemporary theorizations of the complexity of socio-ecological dynamics, by moving away from the traditional understanding of climax ecosystems. Similarly, policies towards the treatment of the bark beetle have been portrayed as a decisive factor for the launch of Czech and German landscapes into the exclusive 1.5 per cent of European protected areas with 'wilderness area' status, and among the few national parks at the global scale with exemplary practices in this regard.

On the other hand, the bark beetle is 'evil', a 'parasite' and a destroyer of historically and culturally valuable forests for the forestry lobby, developers as well as some politicians. This has been particularly true in the case of local leaders based in the South-Bohemian Region, which has administrative jurisdiction over much of the park's territory. They include the current President of the country – Miloš Zeman – who stated that he intended to file a lawsuit against the former environment minister over his failure to prevent the destruction of the park's forests by the bark beetle. One of the key arguments utilized in the context of such debates relates to the economic and employment opportunities afforded by tourism – it has been argued that the landscape's attractiveness for tourism and investment will be detrimentally affected by the 'scarring' caused by the bark beetle infestation and the absence of intact forest cover. However, local people have held an ambivalent attitude towards the rewilding issue, as many of them have supported the activities of environmental organizations that have resisted interventionist efforts to control the bark beetle infestation. The locals' resistance towards the work of the 'chemical brothers' (Stráský and Mánek) is particularly indicative in this regard.

Concluding Remarks

The evidence reviewed in this chapter has highlighted the highly polarized, locally contingent and historically embedded character of environmental policies in Šumava National Park. Debates over current understandings and the future course of nature protection efforts in this area reveal socio-ecological formations and economic relations well beyond the remit of traditional conservation approaches. Particularly indicative in this context are the political relations stemming from the strong attitudes expressed by local people towards the management of the park, possibly due to the presence of a relatively young and well educated population, with a lower-than-average rate of unemployment. Indeed – and very much unlike Pelister – individuals employed in forestry, industry or services dominate the employment structure of the park's residents.

Another key contributing factor are the tensions stemming from Šumava's status as a well-developed tourist destination, faced with continuously rising visitor numbers against the wishes of most of its residents. Of importance here is the strong relationship between the local inhabitants' sense of 'home', on the one hand, and their views of different protected area functions, on the other. This disparity has transpired despite their overall positive appraisal of the existence of the national park, indicating the presence of a pronounced sense of place attachment. Yet the specific socio-demographic structure of the park's population is also reflected in this aspect of its operation: the nature of job opportunities has provided a key residential pull factor for local people. Overall, there is a sense among local people that the national park's existence is having an increasingly beneficial effect on economic growth, mainly thanks to the continued expansion of tourism.

The existence of a close correlation between the local population's perceptions of Šumava National Park as home, versus its functions as a national park and environmental management institution – coupled with their relatively negative appraisal of its nature protection regime – can be attributed to the specific combination of historical developments and current circumstances in the region. Šumava has experienced a series of dramatic demographic, political and economic events in the past, including the designation of its host region as a militarized resource-based periphery at the border of the Iron Curtain for 50 years, accompanied by extensive migration and displacement processes (including the removal of a whole village). Adding to this set of social relations has been the recent dramatic turnaround of the economic and social profile of the region, as a result of the post-communist transition. Furthermore, the management practices of the park have moved from a highly interventionist approach – involving the cutting and removal of vulnerable tree stands – towards a 'natural', non-interventionist strategy aimed at fostering the self-regeneration of the forest.

All of the different historical phases in the evolution of the park's management have prevented the local population from participating in the decision-making process, albeit in different ways. In the first instance, forest exploitation was subcontracted to external companies and took place in areas that were outside the reach of the park's inhabitants, while the latter phase, *inter alia*, has led to the reduction of employment opportunities for local people as full or part-time woodcutters. At the same time, local people still cannot access all parts of the park, as movement paths through the area are strictly defined, and the right of entry into the first zone of 'strict protection' remains prohibited. Local people have also held an ambivalent attitude towards the intense controversies generated by the different viewpoints held by local stakeholders with respect to the management of the bark beetle's impact on the park's forest ecosystems.

It should be noted, however, the National Park Authority's governance policies towards local people have recently started to change. This can mainly be attributed to pressure from experts and environmental non-governmental organizations. Nevertheless, it is clear that the Authority's management practices do not contribute to community-based conservation or co-operative management in an effective manner. The ability of this organization to develop a productive relationship with local communities is further aggravated by numerous structural issues inherited from the past, including the nature of park zoning, past choices of environmental management strategies to deal with the bark beetle infestation, as well as the restitution of historical property to communities. The Authority has been forced to deal with these governance challenges in the face of rising visitor numbers and increased development pressure from national and international tourism. This echoes broader trends in the Czech context, which is well known for experiencing conditions whereby 'more people seek unspoilt landscape settings within a diminishing rural area' thus leading to 'the increasing importance of preserved areas as a recreational hinterland for towns' (Librová 1987).

In spite of the negative perceptions of some of its policies, the Authority has been implementing a selection of measures that are seen favourably by the local population and favour the common interests of all users of the park, mainly with respect to the management of tourist flows. These include the maintenance of cross-country ski tracks, the establishment of 'ecological' buses, the construction of information centres and the rehabilitation of public landmarks in the park (Kušová et al. 2008). But attempts to enforce a rewilding agenda has threatened the success of the national park, leading to the resurfacing of deeper tensions over the future nature protection agendas. Political debates over the adoption of new legislation to regulate Šumava's formal status have allowed such discord to enter the public gaze.

Chapter 8
Conclusions

This book has identified some of the key factors that shape local residents' attitudes and opinions with respect to different national park management systems. I have highlighted how nature protection policies are contingent upon wider environmental, social and political relations, and are heavily influenced by economic opportunities and tourism development patterns. Using evidence drawn from literature reviews, surveys of secondary data, as well as on-site research undertaken in Pelister National Park (Macedonia) and Šumava National Park (Czech Republic), the book has connected the attitudes of local residents towards protected areas as geographical settings for the conduct of everyday life, on the one hand, with the broader issues of public participation and environmental management in the two areas, on the other.

In the text that follows, I outline the main contributions of the book to relevant debates in human geography, anthropology and social environmental science in particular, while summarizing some of the main strands on-site empirical evidence explored in the previous chapters. I also provide a brief discussion of some of the more specific protected area management issues highlighted in the book, particularly with respect to forthcoming research directions and policy measures. The chapter underlines the benefits of studying protected areas in Central and Eastern European countries – both in terms of understanding how the particular legacies of communist central planning have influenced nature conservation, and highlighting some of the global lessons that can be learnt from this region. Concepts discussed in the introductory sections are revisited in this chapter, especially in terms of connecting the consequences of rewilding strategies and aspects of place attachment to local public participation.

The review that I undertook at the beginning of the book found that the conventional management system for protected areas – also known as the 'Yellowstone' model – has led to the displacement of local populations from protected areas, while putting unreasonable limitations on the use of natural resources in the name of nature protection. The reaction to this approach among many scientists and practitioners has often led to the representation of local people as either 'victims' or 'opportunists' with respect to nature conservation. Moving beyond the triangle formed by these two understandings, on the one hand, and the more traditional fortress conservation model, on the other, requires an acknowledgement of the involvement of local people in protected area management as a key necessity for the sustainable and efficient protection of wildlife, and as an economically preferable approach for the effective everyday care and protection of the environment. In cases where it has genuinely succeeded, effective local participation in protected area management has led to the creation of new roles for conservation professionals and protected area authorities, allowing for a fundamental re-framing of existing co-management schemes and inter-institutional arrangements.

It also transpired that the term 'national park' has widely different meanings across the world, while being associated with a diverse range of policy practices. Even though North American (and to a lesser extent, Australian) national parks largely follow the principles of the 'Yellowstone' model, African and South American protected areas that carry this label are largely the outcomes of colonial policies or geopolitical relations. In Asia, national parks were established for conservation purposes beyond traditional environmental reasons; Europe's reserves that fall under this category have been forced to reconcile the complex demands placed by the multiple constituencies they have to serve. Such issues are of particular significance in CEE, where the management of protected areas has taken place against the background of the deep socio-economic and political restructuring recently experienced by the region. The wider reconfiguration of environmental governance in post-socialist countries, therefore, has also been contingent upon the character and direction of nature protection. Yet the multiple social and political aspects of local community participation in the governance of national parks remain insufficiently investigated in the relevant theoretical and policy literature. This is despite the fact that CEE's protected areas are seen by many academics and practitioners as one of Europe's main biodiversity reservoirs and candidates for rewilding policies, not the least due to the lower population densities and higher rates of forest cover seen in this part of the world.

Zooming on to Pelister and Šumava, a number of similarities can be detected at first sight. Both parks are mountainous, forest areas, situated on the border with Greece in the case of the former, and Germany and Austria, in the case of the latter. They lie adjacent to similar protected areas in neighbouring countries. Both of them have been designated at the national level, and are managed by hierarchically organized and rigidly-structured authorities, closely controlled by the respective central governments. In legal terms, Pelister and Šumava alike are Category II protected areas according to International Union for the Conservation of Nature's classification, and as such are meant to contain – formally at least – similar zoning patterns and management practices. Relevant formal legislation in their host countries is also supposed to provide relatively similar background conditions, given that both Macedonia and the Czech Republic have harmonized their environmental protection and nature conservation legislation with that of the EU.

However, the parks differ in terms of their socio-economic history, land-use practices and natural features. There are also numerous social, economic and political nuances in the relationships between the local population and various aspects of protected nature. In essence, the specific development trajectories of the two parks' host regions have combined with the idiosyncrasies of their political, social, economic and institutional circumstances to produce a variegated array of interactions between local people and regulatory practices. This means that attitudes towards the protection, management and impacts of nature protection vary not only among different groups of residents within the parks, but are also specific to particular settlements as well.

One of the main aims of starting this book in the first place – as outlined in the introduction and literature review above – was to provide a more nuanced view of the local population's understandings and experiences of environmental management and nature protection in the two parks. Indeed, my findings challenge the widespread theoretical understanding of local communities in national parks as resistant and passive obstacle in the process of nature protection. In order to corroborate this claim further, I now return to the themes associated with three questions outlined in the introduction (in the section titled 'Purpose of the book').

The Governance of Nature Protection

Both the general literature and the specific findings from Šumava and Pelister have confirmed Goodall and Stabler's (1997) argument that a single model or strategy for nature protection does not exist. And not only: multiple approaches towards environmental governance and management may be in place even in the same protected area. This is particularly evident in Pelister, which is subject to a range of inflexible policy processes that obstruct opportunities for local community participation in protected area management, despite hints of a nascent co-management model. Some of the key factors that contribute to the inflexibility of national park governance and protection in this context include the absence of clear and sustained protected area management strategies, as well the lack of human and financial resources to run local institutions.

The National Park Authority's determination to develop flexible governance solutions that would allow local communities to benefit from the park's existence has, however, helped overcome any potential conflicts over the management of the area. The park has strengthened its relationship with local people thanks to the provision of a wide range of municipal services above and beyond its statutory role. These include waste management, free fuel wood for heating, as well as the improvement and maintenance of local infrastructure (also see Petrova et al. 2009a and 2011). The Authority has also allowed local people to be commercially involved in, and financially benefit from, the trade of non-timber forest products, including *Pinus peuce* seeds and cones, as well as blackberries.

In this context, it is worth noting that the reviewed evidence also indicated that bottom-up modes of local community participation are also present in Pelister to a limited extent. The socio-economic diversity of such practices exceeded my initial expectations, as did their relatively high degree of organization and co-ordination. Taking into account the fact that most protected areas in Macedonia are currently run without a clear strategic approach or genuine understanding about the need to involve local people in the decision-making processes, the existence of these activities in the governance of the national park challenges dominant thinking about the lack of genuine public participation in nature conservation not only in this country, buy the region more widely.

Nature protection and governance issues in Šumava, where locals are significantly more dissatisfied with the work of the national park authority, are markedly different. Governance challenges in this area are significantly more complex and operate on a much larger scale compared to Pelister. In particular, the park has been subject to polarized debates among local people, environmental organizations and state institutions over the implementation of policies underpinned by a 'rewilding' agenda. Adding to this has been the intense pace of tourism growth and economic development, which has resulted in traffic problems and other forms of disturbance that are visibly perceived by the locals. This has not prevented them, however, from retaining a generally positive attitude towards the management of the park and strongly identifying it as their 'home', while developing a favourable opinion about the necessity for its establishment and existence. In addition to the continued threat of increased tourism, however, their underlying objection lies in the rigidity and restrictiveness of the national park's nature protection regime, which limits freedom of movement and employment opportunities.

The State of Environment and Nature in National Parks

Local people may hold negative attitudes toward protected areas, while perceiving them as the cause for a loss of freedom, or as an obstacle to local development (Carrus et al. 2005). However, Neumann (1992) and Brockington et al. (2006) argue that a more relaxed regime of nature protection allows such populations to develop more positive attitudes towards national parks. This counters claims made by Salafsky and Wollenberg (2000) who believe that the consumptive use of biological resources should not be allowed in core zones of the national parks. Thus, moving beyond the understanding of local people as either victims of, or obstacles towards, nature protection has traditionally been seen as a difficult task. A number of authors, however, have come across systematic evidence that moving beyond this dichotomy can be achieved by combining biodiversity conservation and local development goals (see van Schaik and Rijksen 2002, Agrawal and Redford 2009).

As far as the geographical context explored in this book is concerned, the relationship between local people, on the one hand, and nature protection practices and institutions, on the other, reveals a complex and often contradictory picture. Local people in Pelister are generally satisfied with the work of the National Park Authority, while believing that the nature protection regime in the Park is adequate. However, we found many individuals who were unable to evaluate the work of the Authority or the nature protection regime in the Park. Brandon and Dragos (2008) have established that a similar situation exists in Romania's Măcin Mountains National Park. They argue that such circumstances are mainly due to the lack of interaction between the local people and the park. Indeed, Pelister residents found it difficult to express a strong opinion about the quality of their communication with the Park's Authority. Women participating in the survey provided a more positive evaluation of the Authority's work. This is similar to the findings, for example, from Tara National Park in neighbouring Serbia, where female respondents were more positive than males (Tomićević et al. 2010).

Local people in Šumava expressed an overall positive attitude towards the National Park Authority's work as a nature protection organization, although individuals who were satisfied by its work in managing municipal matters were in the minority. However, the residents who evaluated the national park regime as 'strict' were a sizeable group. An almost negligible number of locals were satisfied with the level of communication with the park's Authority, with half of the survey respondents being unaware of the visitors' code. This matches the findings of authors such as Young et al. (2007), who stress that stakeholder participation in Central and Eastern Europe has also been limited to date. Working in a similar vein, Stringer et al. (2006) found that sending a magazine to every household in an Austrian national park was considered a sufficient method of keeping local people involved in management practices.

Both the respondents' level of education and the nature of their jobs were shown to influence the generation of local opinions and attitudes towards the management of Pelister and Šumava alike. This finding is echoed in the broader literature: Petrosillo et al. (2007) stress that the manner in which people perceive environmental quality and sustainability is influenced by, *inter alia*, their socio-economic status, family ties and cultural affiliations (also see Adams 2005, Carrus et al. 2005, Tomićević et al. 2010). Similar trends can be found in CEE: Brandon and Moldovan (2008) reported that local people's attitudes towards the Măcin Mountains National Park in Romania were mostly influenced by the respondents' level of education.

Insufficient employment opportunities are one of the key issues faced by the population of Pelister, where opinions over the job-generating influence of the National Park are divided. These findings are in line with the general socio-economic situation in Macedonia, where unemployment has been a persistent problem in the post-socialist period. They are also indicative of a wider regional situation: Tomićević et al. (2010) stress that the findings from their study of job opportunities in Serbia's Tara National Park were also a reflection of the overall situation in the country. The survey found that approximately 70 per cent of the respondents in Šumava felt that there was a lack of jobs in the Park, while a third thought that the establishment of the park had increased the number of employment opportunities. Respondents with a higher level of education were more negative in these evaluations. Many of my interviewees corroborated findings for Bialowieza National Park, where local residents believed that tourism opportunities are basically the only employment options in the region.

Matsuoka and Kaplan (2008) have emphasized that a significant body of research in the landscape and urban planning domain sees protected nature as a site for improving individuals' quality of life, and providing a sanctuary from 'urban problems'. It follows that nature protection management should, therefore, be aimed at preserving the balance between nature and recreation (see Reynolds and Elson 1996, Cope et al. 1999, Beunen et al. 2008) How this relationship will play out in Pelister remains to be seen: Tourism is still poorly developed in the area at the moment, and the pressure on nature protection and living costs is insignificant. At the same time, Pelister's inhabitants are generally positively inclined towards the increase of tourist and visitor flows in the park, despite having the impression that this has already been happening in recent years.

Conversely, the intense pace of tourism growth and economic development in Šumava National Park has resulted in various forms of disturbance that are visibly perceived by the locals. According to Librová (1987), the Czech context is well known for experiencing conditions in which people seek unspoilt natural settings. Šumava is clearly a well-developed tourist destination, as evidenced by the great number of interviewed residents who point out that the amount of tourists in this park is increasing, as well as the unfavourable attitude of the local population towards further growth in this sector. It is important to note that the income status of tourists in Šumava has undergone changes over time; Cihar and Stankova (2006) note that the number of visitors with a university degree has been on the rise since the 1990s.

Interestingly, 65 per cent of the survey respondents in Šumava did not know about Natura 2000, with half of the individuals within this figure stating that they were insufficiently informed about the framework. These findings echo the experience of other European countries: Papageorgiou and Vogiatzakis (2006) point out that the superimposition of Natura 2000 upon the existing system of protected areas in Greece resulted in the duplication of administrative efforts and related legislation, rendering the overall management of protected areas 'complex, confusing and fragmented' while further complicating the communication with local people and the implementation of participative management practices (476).

Place Attachment

As was pointed out in Chapter 2, I used the concepts of 'ancestors', 'home' and 'migration' as proxies for place attachment in the context of CEE national parks. Placing discussions of nature protection areas within the context of local people's understandings of national parks as 'home' allowed me to evaluate the strength of community affiliations, as well as attitudes towards nature protection and local economic development (Kaltenborn 1998, Kaltenborn et al. 1999, Manzo and Perkins 2006, Petrova et al. 2011).

In their entirety, the multiple strands of evidence explored in this book confirm the validity of Vorkinn and Riese's (2001) work, since they show that place attachment is a key determining variable in the context of local community perceptions of nature protection. Field research of local people's understanding of ancestry, home and migration indicated a strong sense of place attachment in both Šumava and Pelister. This is evidenced by the overwhelming identification of the two parks as 'home', and widespread perceptions that local people would not emigrate from the parks even when given the opportunity. Reflecting a broader situation across Central and Eastern Europe – as well as the work of authors such as Kušová et al. (2008) – ancestral links also played an important role in the lack of long-term residential mobility. The existence of this relationship in Šumava counters national-level discourses which have sought to emphasize the dramatic population movements experienced by the park in its recent past, while seeking to represent its local population as somehow lacking a genuine connection to the area, and a sense of community affiliation (see Furlong 2006).

The homogeneity of local residents' views with respect to questions of place attachment has transpired despite significant socio-demographic differences among them. Broader post-socialist legacies and regional forces may be at play here: work in other protected areas in CEE has also uncovered strong bonds between people and place. Thus, while Tomićić et al. (2010) emphasize the strong place attachment of local people to Tara National Park in Serbia, Adams (2005) stresses that many Sami people would like to live in Swedish urban areas in Sweden, despite possessing legally acknowledged rights to use the natural resources in national parks.

Parallels with neighbouring countries beyond the CEE context can also be drawn. Overall, the socio-demographic characteristics of the respondents in Pelister are similar to the findings from other protected areas in the Balkans such as Serbia, Bulgaria and Greece (see Trakolis 2001, Cellarius 2004, Pavlikakis and Tsihrintzis 2006, Staddon 2009) while the population of Šumava is comparable to those from Central European protected areas such those in Austria, Slovenia, Slovakia, Romania and Poland (see Stringer et al. 2006, Kluvánková-Oravská et al. 2009). The divergent population features of the two national parks are reflected in the driving forces of their residential attractiveness: while Pelister residents cited the desire to live in the countryside as the main reason for staying in the area, it was the nature of employment that provided the key residential pull factor for Šumava.

Both the broader literature and evidence gathered in the field emphasize the existence of a strong relationship between another proxy of place attachment – ancestral links – and local understandings of national parks as organizations with a stewardship role towards the governance of natural heritage. Empirical and conceptual work in this domain has argued that the individuals and local communities who have developed closer socio-economic and cultural bonds to an area display a higher degree of sensitivity to site management and impacts (Stokols and Shumaker 1981, Shumaker and Taylor 1983, Williams and Roggenbuck 1990, Vorkinn and Riese 2001). But some of the evidence from Šumava counters claims made elsewhere – see, for example, Davenport and Anderson (2005) – that local residents can distinguish among specific types of actions towards the management of environmental issues *vis-à-vis* their sense of place attachment. The lack of a correlation between the local population's perceptions of Šumava National Park as home and an institution responsible for nature protection supports such a line of thinking, in addition to residents' relatively negative appraisal of the area's nature protection regime. The specific combination of historical developments and current circumstances in the region may have contributed to this situation, as indicated by the fact that Šumava National Park Authority has struggled to develop successful co-operation mechanisms with the local population.

Although Pelister's residents were strongly attached to the park as their home, they were largely unfamiliar with, or uninterested in, the authority's overall work. In Šumava, local residents were significantly more dissatisfied with the activities of the national park authority. These perceptions have developed against a background context in which the residents of both parks were generally positively inclined towards both the initial designation and the continued presence of the national parks in a broader sense. Many of them saw the areas as places that give them the opportunity to live in an unpolluted, scenic and protected setting, while opening up possibilities for economic investment and job creation. In the wider European context, similar findings have been published by Pavlikakis and Tsihrintzis (2006), Stringer et al. (2006) and Brandon and Dragos (2008).

Furthermore, Šumava residents display a higher awareness of environmental and nature protection issues compared to their Pelister counterparts. This may be, once again, due to micro-scale social, demographic and spatial circumstances: the results from the statistical surveys revealed that the place of residence of the respondents is one of the factors with a key impact on the creation of the attitudes and opinions towards nature protection, management, and tourism development issues. Also influencing these relationships were the respondents' social status and their level of education. Generally, local residents with a higher level of education and with better social status were more critical and opinionated in both parks. This group of respondents also reaped greater benefits from the development of tourism.

Both the literature review and empirical work in the field also pointed to the existence of multiple and overlapping antinomies of community (Watts 2004) with respect to the articulation of notions of belonging among local people in protected areas. While 'communities of place' were clearly in existence in many of the villages that I surveyed in both parks, the areas were also subjected to the distantiated and fluid relations that characterize communities of interest. In both cases, communities demonstrated the ability to act as political subjects capable of shaping the course of nature protection activities.

Limitations

Part of the background research that fed into this book was part of a wider research project, whose programme was influenced by the aims and objectives of my own work. In particular, the questionnaire used in my survey had to be analogous to the one used in the project, in order to allow for cross-comparisons. This hampered the efficacy of the data gathering process, particularly due to the fact that the questionnaire is rather long and time consuming.

Also, national parks are complex ecological-social systems that require investigations from a range of different perspectives. Owing to time and resource limitations, my research was mainly focused on the broader literature on the subject, policy frameworks and local residents' views. Less attention was paid to representatives of state organizations, municipal bodies, private businesses and national park management authorities. Adding to these limitations is the fact that Šumava National Park is a large area with a dispersed distribution of people. As a result, the research was undertaken only in the central part of the park. The specific geographical focus of the study can also be attributed to the fact that the same area was part of previous research, and as such provided an easier time frame for comparison.

Moving On ...

The main policy implications of this study revolve around the need for improved education and employment opportunities for local people, as key ways towards achieving improved national park management across the world. Also, I would emphasize the need to support national institutions in the 'transition' context of CEE towards developing more effective governance practices. This could be achieved via the involvement of local people in biodiversity monitoring, the dissemination of relevant information to stakeholders, as well as educational courses for training national park residents to make use of local economic opportunities. Increased education and awareness-raising programmes can also help, especially if they involve sustained input from the scientific community. Mechanisms for addressing issues at the local people–visitor nexus may entail the engagement of tourists in the monitoring of species (e.g. bird watching), and obtaining tourist donations or support for nature and cultural heritage projects. The resolution of tensions surrounding the implementation of rewilding agendas necessitates the development of effective compensation mechanisms in cases where protected nature and wildlife has damaged the property or livelihoods of local residents.

In terms of some of the specific lessons from the case study context, it can be said that there is a clear need for a more open communication with local authorities and a greater degree of participation in the articulation of existing policies. Nevertheless, it is worth noting that the National Park Authority in Pelister has managed to develop a relatively successful economic co-operation dynamic with local people. This is despite the residents' lack of information about, and insufficient involvement in, the decision making and management practices associated the park. But the future implementation of the park's management programmes will require significant financial, technical and human resources as well as an effective participatory approach in order to succeed. Otherwise there is a danger that the sustainable development of the area may be substantially threatened, especially in light of the fact that both tourist numbers and the local residents' level of dissatisfaction with the management of the park are already on the increase.

Overall, the reviewed evidence challenges the view that local populations and communities are obstacles towards the sustainable management of national parks, as a result of their allegedly negative attitude towards nature conservation and protection. The local populations included in this study maintain a positive attitude towards the establishment of national parks, while possessing a diverse set of views about the different aspects of their operation. The main reason for such opinions lies in the residents' belief that protected areas provide opportunities for greater economic investment and growth. Also, local populations cannot be treated as a monolith in terms of their views of national park management, which are not necessarily correlated to perceptions of place attachment. The residents of national parks are able to distinguish between the different nature protection functions of protected areas and their governance, while providing responses that vary significantly according to location, education and social status. This finding is far from the widespread theoretical understanding of local communities in national parks as uniform and passive barriers in the process of nature protection.

Future research endeavours could push the boundaries of existing understandings of the socio-spatial differences in the perception of home, belonging and place within protected areas, while establishing the role of different participation models in creating such differences. With its ability to provide a comprehensive conceptual and epistemological framework involving sophisticated theorizations of place, nature and human behaviour, I would argue, social environmental science is in an ideal position to play a central role in this undertaking.

References

Aagesen, D. 2000. Rights to land and resources in Argentina's Alerces National Park. *Bulletin of Latin American Research* 19: 547–569.

Adams, J.S. 1992. *The Myth of Wild Africa: Conservation without Illusion.* Berkeley: University of California Press.

Adams, M. 2005. Beyond Yellowstone? Conservation and Indigenous rights in Australia and Sweden. In *Discourses and Silences: Indigenous Peoples, Risks and Resistance*, edited by G. Cant, G.A. and J. Inns. Christchurch, New Zealand: Department of Geography, University of Canterbury. 127–138.

Adams, W.M. 2003. *Future Nature: A Vision for Conservation.* London: Earthscan.

Adams, W.M. and Hulme, D. 2001. If community conservation is the answer in Africa, what is the question? *Oryx* 35: 193–200.

Adger, W.N. 2001. Scales of governance and environmental justice for adaptation and mitigation of climate change. *Journal of International Development* 13: 921–931.

Adger, W.N. 2010. Social Capital, Collective Action, and Adaptation to Climate Change. In *Der Klimawandel*, edited by M. Voss. VS Verlag für Sozialwissenschaften. 327–345, http://link.springer.com/chapter/10.1007/978-3-53 1-92258-4_19 (accessed January 4, 2014).

Agrawal, A. and Gibson, C.C. 1999. Enchantment and Disenchantment: The Role of Community in Natural Resource Conservation. *World Development* 27: 629–649.

Agrawal, A. and Ostrom, E. 2006. Political science and conservation biology: A dialog of the deaf. *Conservation Biology* 20: 681–682.

Agrawal, A. and Redford, K. 2006. *Poverty, Development, and Biodiversity Conservation: Shooting in the Dark?* Ann Arbor, MI: WCS Working Paper.

Agrawal, A. and Redford, K. 2009. Conservation and displacement: An overview. *Conservation and Society* 7: 1–10.

Agyeman, J. 1995. Environment, heritage and multiculturalism. *Journal of the Association of Heritage Interpretation* 1: 5–6.

Agyeman, J. and Spooner, R. 1997. Ethnicity and the rural environment. In *Contested Countryside Cultures: Otherness, Marginalisation and Rurality*, edited by P. Cloke and J. Little. London: Routledge. 197–217.

Ale, S.B. and Howe, H.F. 2010. What Do Ecological Paradigms Offer to Conservation? *International Journal of Ecology* 2010, http://downloads. hindawi.com/journals/ijeco/2010/250754.pdf.

Altman, I. and Low, S.M. 1992. *Place Attachment.* New York and London: Plenum Press.

Amend, S. and Amend, T. 1995. *National Parks without People? The South American Experience*. IUCN – The World Conservation Union.

Antolini, D.E. 2008. National Park Law in the U.S.: Conservation, Conflict, and Centennial Values. *William & Mary Environmental Law and Policy Review* 33: 851.

Antrop, M. 2001. The language of landscape ecologists and planners: A comparative content analysis of concepts used in landscape ecology. *Landscape and Urban Planning* 55: 163–173.

Baird, I. 1999. Fishing for sustainability in the Mekong Basin Watershed. *TERRA* 4: 54–56.

Baird, I. 2000. Integrating Community-Based Fisheries Co-Management and Protected Areas Management in Lao PDR: Opportunities for Advancement and Obstacles to Implementation, Evaluating Eden Series. *International Institute for Environment and Development: Discussion Paper No. 14*.

Bajracharya, S.B. and Dahal, N. 2008. *Shifting Paradigms in Protected Area Management*. Kathmandu: National Trust for Nature Conservation.

Baker, S. and Jehlička, P. 1998. Dilemmas of transition: The environment, democracy and economic reform in East Central Europe – an introduction, http://www.tandfonline.com/doi/pdf/10.1080/09644019808414370 (accessed January 6, 2014).

Bakker, K. 2010. The limits of 'neoliberal natures': Debating green neoliberalism. *Progress in Human Geography* 34: 715–735.

Balakrishnan, M. and Ndhlovu, D.E. 1992. Wildlife utilization and local people: A case-study in Upper Lupande Game Management Area, Zambia. *Environmental Conservation* 19: 135–144.

Báldi, A. and Faragó, S. 2007. Long-term changes of farmland game populations in a post-socialist country (Hungary). *Agriculture, Ecosystems & Environment* 118: 307–311.

Barnett, C. 2005. The consolations of 'neoliberalism'. *Geoforum* 36: 7–12.

Baron, J.S., Gunderson, L., Allen, C.D., et al. 2009. Options for National Parks and Reserves for Adapting to Climate Change. *Environmental Management* 44: 1033–1042.

Barraclough, S.L. 2001. *Toward Integrated and Sustainable Development?* United Nations Research Institute for Social Development.

Barrow, E. and Murphree, M. 2001. Community conservation: From concept to practice. In *African Wildlife and Livelihoods: The Promise and Performance of Community Conservation*, edited by D. Hulme and M. Murphree. London: James Curry.

Bartlein, P.J., Whitlock, C. and Shafer, S.L. 1997. Future Climate in the Yellowstone National Park Region and Its Potential Impact on Vegetation. *Conservation Biology* 11: 782–792.

Bawa, K.S., Seidler, R. and Raven, P.H. 2004. Reconciling Conservation Paradigms. *Conservation Biology* 18: 859–860.

Bell, C. and Lyall, J. 2002. *The Accelerated Sublime: Landscape, Tourism, and Identity*. Connecticut: Greenwood Publishing Group.

Berkes, F. 2004. Rethinking community-based conservation. *Conservation Biology* 18: 621–630.

Berkes, F., Colding, J.F. and Folke, C. 2003. *Navigating Nature's Dynamics: Building Resilience for Complexity and Change*. New York: Cambridge University Press.

Beunen, R., Regnerus, H.D. and Jaarsma, C.F. 2008. Gateways as a means of visitor management in national parks and protected areas. *Tourism Management* 29: 138–145.

Blaikie, P. and Jeanrenaud, S. 1997. Biodiversity and Human Welfare. In *Social Change and Conservation: Environmental Politics and Impacts of National Parks and Protected Areas*, edited by K.B. Ghmire and M.P. Pimbert. London: Earthscan.

Blunt, A. and Dowling, R. 2006. *Home*. London: Routledge.

Booth, K.L. and Simmons, D.G. 2000. Tourism and the establishment of national parks in New Zealand. 39–49.

Botcheva, L. 2001. Expertise and international governance: Eastern Europe and the adoption of European Union environmental legislation. *Global Governance* 7: 197–224.

Bourdieu, P. 1986. Forms of capital. In *Handbook of Theory and Research in the Sociology of Education*, edited by J. Richardson. New York: Greenwood Press. 241–258.

Bouzarovski, S. 2009. East-Central Europe's changing energy landscapes: A place for geography. *Area* 41: 452–463.

Bowman, M. and Hunter, D. 1991. Environmental Reforms in Post-Communist Central Europe: From High Hopes to Hard Reality. *Michigan Journal of International Law* 13: 921–980.

Brandon, A. and Dragos, M. 2008. Poised for engagement? Local communities and Măcin Mountains National Park, Romania. *The International Journal of Biodiversity Science and Management* 4: 230–241.

Brandon, K., Redford, K.H. and Sanderson, S.E. 1998. *Parks in Peril: People, Politics, and Protected Areas*. Washington, DC: The Nature Conservancy.

Breakell, B. 2002. Missing persons: Who doesn't visit the people's parks? *Countryside Recreation* 10: 13–17.

Brockington, D. 2004. Book review: Contested nature: Promoting international biodiversity with social justice in the twenty-first century (eds), Brechin, S., Wilhuse, P., Fortwangler, C. and West, P. *Journal of Ecological Anthropology* 8: 84–85.

Brockington, D. and Igoe, J. 2006. Eviction for conservation: A global overview. *Conservation and Society* 4: 424–470.

Brockington, D., Duffy, R. and Igoe, J. 2008. *Nature Unbound: Conservation, Capitalism and The Future of Protected Areas*. London: Earthscan.

Brockington, D., Igoe, J. and Schmidt-Soltau, K. 2006. Conservation, human rights, and poverty reduction. *Conservation Biology* 20: 250–252.

Brosius, J.P., Tsing, A.L. and Zerner, C. 1998. Representing communities: Histories and politics of community-based natural resource management. *Society & Natural Resources* 11: 157–168.

Brovko, P.F. and Fomina, N.I. 2008. The history of establishment of the national park network in countries of the Asian-Pacific region. *Geography and Natural Resources* 29: 221–225.

Browning, C.S. 2003. The region-building approach revisited: The continued othering of Russia in discourses of region-building in the European North. *Geopolitics* 8: 45–71.

Buckley, R. 2003. Pay to Play in Parks: An Australian Policy Perspective on Visitor Fees in Public Protected Areas. *Journal of Sustainable Tourism* 11: 56–73.

Bursik, R.J. 1988. Social disorganization and theories of crime and delinquency: Problems and prospects. *Criminology* 26: 519–552.

Büscher, B. and Dressler, W. 2012. Commodity conservation. The restructuring of community conservation in South Africa and the Philippines. *Geoforum* 43: 377–376.

Buzarovski, S. 2001. Local environmental action plans and the 'glocalisation' of post-socialist governance. *GeoJournal* 55: 557–568.

Carmin, J. and Fagan, A. 2010. Environmental mobilisation and organisations in post-socialist Europe and the former Soviet Union. *Environmental Politics* 19: 689–707.

Carmin, J. and Vandeveer, S.D. 2004. Enlarging EU Environments: Central and Eastern Europe from Transition to Accession. *Environmental Politics* 13: 3–24.

Carrus, G., Bonaiuto, M. and Bonnes, M. 2005. Environmental concern, regional identity, and support for protected areas in Italy. *Environment and Behavior* 37: 237–257.

Castán Broto, V., Burningham, K., Carter, C. and Elghali, L. 2010. Stigma and attachment: performance of identity in an environmentally degraded place. *Society & Natural Resources* 23: 1–17.

Castree, N. 2011. Neoliberalism and the Biophysical Environment 3: Putting theory into practice. *Geography Compass* 5: 35–49.

Castree, N. 2013. *Making Sense of Nature*. Routledge.

Castree, N. 2007. Neoliberalising nature: Processes, effects, and evaluations. *Environment and Planning A* 40: 153–173.

Castro, P.A. and Nielson, E. 2004. Natural resource conflict management case studies: An analysis of power, participation and protected areas. *Forest Ecology and Management* 193: 427–428.

Cellarius, B.A. 2004. *In the Land of Orpheus: Rural Livelihoods and Nature Conservation in Postsocialist Bulgaria*. Madison: University of Wisconsin Press.

Ch'oc, G. 2012. Maya in Belize Hope to Set Historic FIP Settlement. *Indigenous Policy Journal* 23, http://www.indigenouspolicy.org/index.php/ipj/article/view/130 (accessed January 6, 2014).

Chape, S., Harrison, J., Spalding, M. and Lysenko, I. 2005. Measuring the extent and effectiveness of protected areas as an indicator for meeting global biodiversity targets. *Philosophical Transactions of the Royal Society B: Biological Sciences* 360: 443–455.

Chape, S., Spalding, M. and Jenkins, M. 2008. *The World's Protected Areas: Status, Values and Prospects in the 21st Century*. Spain: Universidad de Castilla La Mancha.

Child, B. and Barnes, G. 2010. The conceptual evolution and practice of community-based natural resource management in southern Africa: Past, present and future. *Environmental Conservation* 37: 283–295.

Cihar, M. and Stankova, J. 2006. Attitudes of stakeholders towards the Podyji/Thaya River Basin National Park in the Czech Republic. *Journal of Environmental Management* 81: 273–285.

Cihar, M., Tancosova, Z. and Trebicky, V. 2000. Selected aspects of the sustainable development in Šumava National Park as seen by the local residents. *Silva Gabreta* 5: 195–216 (in Czech).

Cihar, M., Trebicky, V. and Novak, J. 2001. Selected indicators of sustainable tourism in the central part of the Sumava National Park and Biosphere Reserve. *Silva Gabreta* 6: 295–304.

Clapp, R.A. 1998. Regions of refuge and the agrarian question: Peasant agriculture and plantation forestry in Chilean araucanía. *World Development* 26: 571–589.

Clark, M.L. and Aide, T.M. 2011. An analysis of decadal land change in Latin America and the Caribbean mapped from 250-m MODIS data. In *34th International Symposium on Remote Sensing of Environment, Sydney, Australia*. 10–15, http://www.isprs.org/proceedings/2011/ISRSE-34/211104015Final00711.pdf (accessed January 6, 2014).

Clough, P.W.J. and Meister, A.D. 1989. Benefit assessment of recreational land: The Whakapapa area, Tongariro National Park. 89.

Colchester, M. 1997. Salvaging nature: Indigenous peoples and protected areas. In *Social Change and Conservation: Environmental Politics and Impacts of National Parks and Protected Areas*, edited by K.B. Ghimire and M.P. Pimbert. London: Earthscan. 97–130.

Colchester, M. 2004. Conservation policy and indigenous people. *Environmental Science and Policy* 7: 145–153.

Coleman, J.S. 1988. Social Capital in the Creation of Human Capital. *The American Journal of Sociology* 94.

Cope, A., Doxford, D. and Millar, G. 1999. Counting users of informal recreation facilities. *Managing Leisure* 4: 229–244.

Crompton, J.L. 1998. Forces Underlying the Emergence of Privatization in Parks and Recreation. *Journal of Park and Recreation Administration* 16, http://js.sagamorepub.com/jpra/article/view/1669 (accessed January 6, 2014).

Crompton, J.L. and Lamb, C.W. 1986. *Marketing Government and Social Services*. New Jersey: Wiley.

Cronon, W. 1995. The trouble with wilderness; or, getting back to the wrong nature. In *Uncommon Ground: Rethinking the Human Place in Nature*, edited by W. Cronon. New York: W.W. Norton and Co.

Cross, M.S., Zavaleta, E.S., Bachelet, D., et al. 2012. The Adaptation for Conservation Targets (ACT) framework: A tool for incorporating climate change into natural resource management. *Environmental Management* 50: 341–351.

Cuba, L. and Hummon, D.M. 1993. A place to call home: Identification with dwelling, community and region. *The Sociological Quarterly* 34: 111–131.

Cundill, G., Thondhlana, G., Sisitka, L., et al. 2013. Land claims and the pursuit of co-management on four protected areas in South Africa. *Land Use Policy* 35: 171–178.

Daily, G. and Ellison, K. 2002. *The New Economy of Nature: The Quest to Make Conservation Profitable*. Washington, DC: Island Press.

Davenport, M.A. and Anderson, D.H. 2005. Getting from sense of place to place-based management: An interpretive investigation of place meanings and perceptions of landscape change. *Society & Natural Resources* 18: 625–641.

Davis, C.R. and Hansen, A.J. 2011. Trajectories in land use change around U.S. National Parks and challenges and opportunities for management. *Ecological Applications* 21: 3299–3316.

De Castro, F., Siqueira, A.D., Brondízio, E.S. and Ferreira, L.C. 2006. Use and misuse of the concepts of tradition and property rights in the conservation of natural resources in the Atlantic Forest (Brazil). *Ambiente e Sociedade* 9.

De Sans, À.P. 2004. Sense of place and migration histories. Idiotopy and idiotope. *Area* 36: 348–357.

De Sherbinin, A. 2008. Is poverty more acute near parks? An assessment of infant mortality rates around protected areas in developing countries. *Oryx* 42: 26.

Defries, R., Hansen, A., Turner, B.L., et al. 2007. Land use change around protected areas: Management to balance human needs and ecological function. *Ecological Applications* 17: 1031–1038.

Delanty, G. 2013. *Community*. London: Routledge.

Dey, C. 1997. Women, forest products and protected areas: A case study of Jaldapara Wildlife Sanctuary, West Bengal, India. In *Social Change and Conservation: Environmental Politics and Impacts od National Parks and Protected Areas*, edited by K.B. Ghimire and M.P. Pimbert. London: Earthscan.

Diamond, J. 1995. Easter's End. *Discover* 16, http://thechemguy.com/APchem/Easter%20Island.pdf.

Diemer, M., Held, M. and Hofmeister, S. 2003. Urban wilderness in Central Europe: Rewilding at the urban fringe. *International Journal of Wilderness* 9: 7–11.

Disman, M. 1993. *Jak se vyrábí sociologická znalost*. Praha: Vydavatelství Karolinum Univerzita Karlova.

Dowie, M. 2009. *Conservation Refugees: The Hundred-Year Conflict between Global Conservation and Native People*. Cambridge, MA and London: MIT Press.

Dudley, N. 2008. *Guidelines for Applying Protected Area Management Categories*. IUCN.

Duffield, J.W., Neher, C.J. and Patterson, D.A. 2008. Wolf recovery in Yellowstone: Park visitor attitudes, expenditures, and economic impacts. *Yellowstone Science* 1: 20–25.

Duffy, R. 2006. The potential and pitfalls of global environmental governance: The politics of transfrontier conservation areas in Southern Africa. *Political Geography* 25: 89–112.

Duffy, R. 2010. *Nature Crime: How We're Getting Conservation Wrong*. London: Yale University Press.

Duffy, R. and Moore, L. 2010. Neoliberalising Nature? Elephant-back tourism in Thailand and Botswana. *Antipode* 42: 742–766.

Dunn, K. C. 2009. Contested State Spaces: African National Parks and the State. *European Journal of International Relations* 15: 423–446.

Durkheim, E. 1973. *Emile Durkheim on Morality and Society*. University of Chicago Press.

Eagles, P.F.J. and McCool, S.F. 2002. *Tourism in National Parks and Protected Areas: Planning and Management*. Wallingford: CABI Publishing.

EAR (European Agency for Reconstruction). 2003. *Pelister Tourism Development Study: Development Concept Report*. Thessaloniki: European Agency for Reconstruction.

Edensor, T. 2000. Walking in the British countryside: Reflexivity, embodied practices and ways to escape. *Body and Society* 6: 81–106.

Edmonds, R.L. 2002. *Patterns of China's Lost Harmony: A Survey of the Country's Environmental Degradation and Protection*. London: Routledge.

Eriksson, M., Lilja, S. and Roininen, H. 2006. Dead wood creation and restoration burning: Implications for bark beetles and beetle induced tree deaths. *Forest Ecology Management* 231: 205–213.

Escobar, A. 2011. *Encountering Development: The Making and Unmaking of the Third World*. New Jersey: Princeton University Press.

Etzioni, A. 1995. *The Responsive Community: A Communitarian Perspective*. Rochester, NY: Social Science Research Network, http://papers.ssrn.com/abstract=1437135 (accessed January 4, 2014).

European Commission. 1996. *Council Decision on the Principles, Priorities and Conditions contained in the European Partnership with the Former Yugoslav Republic of Macedonia*. Brussels: European Commission.

European Commission. 2005. *Council Decision on the Principles, Priorities and Conditions contained in the European Partnership with the Former Yugoslav Republic of Macedonia*. Brussels: European Commission.

European Commission. 2013. *Guidelines on Wilderness in Natura 2000: Management of Terrestrial Wilderness and Wild Areas within the Natura 2000 Network*. Brussels: European Commission.

Featherstone, M. 1991. *Consumer Culture and Postmodernism*. London: Sage.

Ferguson, J. 2006. *Global Shadows: Africa in the Neoliberal World Order*. Duke University Press.

Field, J. 2003. *Social Capital*. East Sussex: Psychology Press.

Folke, C., Carpenter, S., Elmqvist, T., et al. 2002. Resilience and sustainable development: Building adaptive capacity in a world of transformations. *Ambio* 31: 437–440.

Franklin, S. 2002. Bialowieza Forest, Poland: Representation, myth, and the politics of dispossession. *Environment and planning A* 34: 1459–1486.

Furlong, K. 2006. Unexpected narratives in conservation: Discourses of identity and place in Šumava National Park, Czech Republic. *Space and Policy* 10: 47–65.

Gandiwa, E., Heitkönig, I.M., Lokhorst, A.M., et al. 2013. CAMPFIRE and human-wildlife conflicts in local communities bordering northern Gonarezhou National Park, Zimbabwe. *Ecology and Society* 18: 7.

Ghimire, K.B. and Pimbert, M.P. 1997a. Preface. In *Social Change and Conservation: Environmental Politics and Impacts of National Parks and Protected Areas*, edited by K. B. Ghimire and M.P. Pimbert. London: Earthscan.

Ghimire, K.B. and Pimbert, M.P. 1997b. Social change and conservation: An overview of issues and concepts. In *Social Change and Conservation: Environmental Politics and Impacts of National Parks and Protected Areas*, edited by K.B. Ghimire and M.P. Pimbert. London: Earthscan. 1–45.

Gladden, J.N. 2001. Arctic wilderness policy in the United States and Finland. *Environmental Management* 27: 367–376.

Gläser, J. 2001. 'Producing Communities' as a Theoretical Challenge. *Paper presented at the Australian Sociological Association Conference*.

Goldman, M. 1998. *Privatizing Nature: Political Struggles for The Global Commons*. London: Pluto Press.

Goodall, B. and Stabler, M. 2000. Environmental standards and performance measurement in tourism destination development. In *Tourism and Sustainable Community Development*, edited by G. Richards and D. Hall. London and New York: Routledge. 63–82.

Goodall, B. and Stabler, M.J. 1997. Principles influencing the determination of environmental standards for sustainable tourism. In *Tourism and Sustainability: Principles to Practice*, edited by M.J. Stabler. Wallingford: CAB International.

Goodwin, H. 1996. In pursuit of ecotourism. *Biodiversity and Conservation* 5: 277–291.

Gössling, S. 1999. Ecotourism: A means to safeguard biodiversity and ecosystem functions? *Ecological Economics* 29: 303–320.

Gray, S. and McCabe, T. 2010. Conservation Efforts, National Parks and the Indigenous Hill Tribes of Northern Thailand: An Examination of the Region's Historical Conflicts and Current Rights-Based Approaches, http://www.

consortiumjournal.com/c2011/content/Gray,%20Shalana--Thailand%20 Conservation.pdf (accessed January 6, 2014).

GRM (Government of the Republic of Macedonia). 1993. *Act on Fishing*. Skopje: Official Gazette of the Republic of Macedonia.

GRM (Government of the Republic of Macedonia). 2004a. *Act on Hunting*. Skopje: Official Gazette of the Republic of Macedonia.

GRM (Government of the Republic of Macedonia). 2004b. *Act on Nature Protection*. Skopje: Official Gazette of the Republic of Macedonia.

GRM (Government of the Republic of Macedonia). 2011. *Act on Plant Protection*. Skopje: Official Gazette of the Republic of Macedonia.

GRM (Government of the Republic of Macedonia). 2013. *Act on Forestry*. Skopje: Official Gazette of the Republic of Macedonia.

Grodzki, W., Mcmanus, M., Knizek, M., et al. 2004. Occurrence of spruce bark beetles in forest stands at different levels of air pollution stress. *Environmental Pollution* 130: 73–83.

Grove, R.H. 1995. *Green Imperialism: Colonial Expansion, Tropical Island Edens and the Origins of Environmentalism, 1600–1860*. Cambridge: Cambridge University Press.

Grujičić, I., Milijić, V. and Nonić, D. 2008. Conflict management in protected areas: The Lazar Canyon natural monument, Eastern Serbia. *International Journal of Biodiversity Science, Ecosystem Services & Management* 4: 219–229.

Hall, C.M. 1999. Rethinking collaboration and partnership: A public policy perspective. *Journal of Sustainable Tourism* 7: 274–289.

Hall, D. 2000. Identity, community and sustainability: Prospects for rural tourism in Albania. In *Tourism and Sustainable Community Development*, edited by G. Richards and D. Hall. London and New York: Routledge. 48–59.

Hansen, A.J., Davis, C.R., Piekielek, N., et al. 2011. Delineating the ecosystems containing protected areas for monitoring and management. *BioScience* 61: 363–373.

Hartig, T. 1993. Nature experience in transactional perspective. *Landscape and Urban Planning* 25: 17–36.

Harvey, D. 2007. *A Brief History of Neoliberalism*. Oxford: Oxford University Press.

Heller, N.E. and Zavaleta, E.S. 2009. Biodiversity management in the face of climate change: A review of 22 years of recommendations. *Biological Conservation* 142: 14–32.

Herath, G. 2002. Research methodologies for planning ecotourism and nature conservation. *Tourism Economics* 8: 77–101.

Herzog, T.R. 1989. A cognitive analysis of preference for urban nature. *Journal of Environmental Psychology* 9: 27–43.

Hewett, T. and Fletcher, S. 2009. The emergence of service-based integrated coastal management in the UK. *Area* 42: 313–327.

Heynen, N. and Robbins, P. 2005. The neoliberalization of nature: Governance, privatization, enclosure and valuation. *Capitalism, Nature, Socialism* 16: 5–8.

Hidalgo, M.C. and Hernández, B. 2001. Place attachment: Conceptual and empirical questions. *Journal of Environmental Psychology* 21: 273–281.

Hillstrom, K. and Hillstrom, L.C. 2004. *Latin America and the Caribbean: A Continental Overview of Environmental Issues*. California: ABC-CLIO.

Hiwasaki, L. 2005. Toward sustainable management of national parks in Japan: Securing local community and stakeholder participation. *Environmental Management* 35: 753–764.

Hlásny, T., Zajívcková, L., Turvcáni, M., et al. 2011. Geographical variability of spruce bark beetle development under climate change in the Czech Republic. *Journal of Forest Science* 57: 242–249.

Höchtl, F., Lehringer, S. and Konold, W. 2005. 'Wilderness': What it means when it becomes a reality – a case study from the southwestern Alps. *Landscape and Urban Planning* 70: 85–95.

Hodgson, D.L. 2002. Precarious alliances: The cultural politics and structural predicaments of the Indigenous rights movement in Tanzania. *American Anthropologist* 104: 1086–1097.

Hogan, Z. 1997. Aquatic conservation zones: Community management of rivers and fisheries. *TERRA* 3: 29–33.

Hopkins, J.W. 1995. *Policymaking for Conservation in Latin America: National Parks, Reserves, and the Environment*. Conneticut: Greenwood Publishing Group.

Hough, J.L. 1988. Obstacles to effective management of conflicts between national parks and surrounding human communities in developing countries. *Environmental Conservation* 15: 129–136.

Hubbard, P. and Holloway, L. 2000. *People and Place: The Extraordinary Geography of Everyday Life*. Harlow: Prentice Hall.

Hughes, M. and Carlsen, J. 2011. National park user pays systems in Australia. Cost recovery vs access for all? *Journal of Tourism and Leisure Studies* 17: 129–146.

Hulme, D. and Murphree, M. 2001. *African Wildlife and Livelihoods: The Promise and Performance of Community Conservation*. London: James Curry.

Hummon, D.M. 1990. *Commonplaces: Community Ideology and Identity in American Culture*. Albany: State University of New York Press.

Hutton, J., Adams, W.M. and Murombedzi, J.C. 2005. Back to the barriers? Changing narratives in biodiversity conservation. *Forum for Development Studies* 32: 341–370.

Hutton, S., Kolpeja, V., Novkovska, B., et al. 2000. Albania and Macedonia: transitions and poverty. In *Poverty in Transition*, edited by S. Hutton. London: Routledge. 125–139.

Igoe, J. 2010. Shifting Paradigms, Indigeneity, and Neoliberal Conservation in Tanzania's Northern Tourist Circuit. Paper presented at the Annual Meeting, Washington, DC, 16 April 2010.

Igoe, J. and Brockington, D. 2007. Neoliberal conservation: A brief introduction. *Conservation and Society* 5: 432–449.

Igoe, J., Neves, K. and Brockington, D. 2010. A spectacular eco-tour around the historic bloc: Theorising the convergence of biodiversity conservation and capitalist expansion. *Antipode* 42: 486–512.

Illner, M. 1999. Territorial decentralization: an obstacle to democratic reform in central and eastern Europe. *The Transfer of Power: Decentralization in Central and Eastern Europe* (Budapest: Local Government and Public Service Reform Initiative), 7–42.

Iojă, C.I., Pătroescu, M., Rozylowicz, L., et al. 2010. The efficacy of Romania's protected areas network in conserving biodiversity. *Biological Conservation* 143: 2468–2476.

IUCN. 1969. *Standards and Nomenclature for Protected Areas: Resolution from the 10th General Assembly of IUCN in New Delhi, India.* New Delhi: IUCN.

IUCN. 2010. Protected Area Management Categories, http://www.unep–wcmc. org/protected_areas/categories/index.html.

Jentoft, S., van Son, T.C. and Bjørkan, M. 2007. Marine Protected Areas: A governance system analysis. *Human Ecology* 35: 611–622.

Jepson, P. and Whittaker, R.J. 2002. Histories of protected areas: Internationalisation of conservationist values and their adoption in the Netherlands Indies (Indonesia). *Environment and History* 8: 129–172.

Job, H. 2008. Estimating the Regional Economic Impact of Tourism to National Parks: Two Case Studies from Germany. *GAIA – Ecological Perspectives for Science and Society* 17: 134–142.

Jonášová, M. and Prach, K. 2008. The influence of bark beetles outbreak vs. salvage logging on ground layer vegetation in Central European mountain spruce forests. *Biological Conservation* 141: 1525–1535.

Jönsson, A.M., Appelberg, G., Harding, S. and Bärring, L. 2009. Spatio-temporal impact of climate change on the activity and voltinism of the spruce bark beetle, Ips typographus. *Global Change Biology* 15: 486–499.

Jump, A.S., Mátyás, C. and Peñuelas, J. 2009. The altitude-for-latitude disparity in the range retractions of woody species. *Trends in Ecology & Evolution* 24: 694–701.

Kaltenborn, B.P. 1998a. Effects of sense of place on responses to environmental impacts. *Applied Geography* 18: 169–189.

Kaltenborn, B.P. 1998b. Effects of sense of place on responses to environmental impacts: A study among residents in Svalbard in the Norwegian high Arctic. *Applied Geography* 18: 169–189.

Kaltenborn, B.P., Riese, H. and Hundeide, M. 1999. National park planning and local participation: Some reflections from a mountain region in Southern Norway. *Research and Development* 19: 51–61.

Kaltenborn, B.P., Vistad, O.I. and Stanaitis, S. 2002. National parks in Lithuania: Old environment in a new democracy. *Norsk Geografisk Tidsskrift – Norwegian Journal of Geography* 56: 32–40.

Kapoor, I. 2001. Towards participatory environmental management? *Journal of Environmental Management* 63: 269–279.

Keenleyside, C. and Tucker, G.M. 2010. *Farmland Abandonment in the EU: an Assessment of Trends and Prospects*. Report prepared for WWF. London: Institute for European Environmental Policy.

Kellert, S.R., Mehta, J.N., Ebbin, S.A. and Lichtenfeld, L.L. 2000. Community natural resource management: Promise, rhetoric, and reality. *Society and Natural Resources* 13: 705–715.

Kelly, L.A. 1995. *Race and place: Denning community in the post-Shaw era*. Paper presented at the Annual Meeting of the Association of American Geographers.

Kilpatrick, S., Field, J. and Falk, I. 2003. Social capital: An analytical tool for exploring lifelong learning and community development. *British Educational Research Journal* 29: 417–433.

King, B. 2010. Conservation geographies in Sub-Saharan Africa: The politics of National Parks, community conservation and peace parks. *Geography Compass* 4: 14–27.

Kluvánková-Oravská, T., Chobotova, V., Banaszak, I., et al. 2009. From government to governance for biodiversity: The perspective of Central and Eastern European transition countries. *Environmental Policy and Governance* 19: 186–196.

Knott, C.H. 1998. *Living with The Adirondack Forest: Local Perspectives on Land Use Conflicts*. New York: Cornell University Press.

Kornai, J. 1992. The Postsocialist Transition and the State: Reflections in the Light of Hungarian Fiscal Problems. *American Economic Review* 82: 1–21.

Kramer, R.A. 1997. *Last Stand: Protected Areas and the Defense of Tropical Biodiversity*. Oxford: Oxford University Press.

Krenova, Z. and Kiener, H. 2012. Europe's Wild Heart – still beating? Experiences from a new transboundary wilderness area in the middle of the Old Continent. *European Journal of Environmental Sciences* 2, http://www.ejes.cz/index.php/ejes/article/view/102 (accessed January 7, 2014).

Kuemmerle, T., Hostert, P., Radeloff, V.C., et al. 2007. Post-socialist forest disturbance in the Carpathian border region of Poland, Slovakia, and Ukraine. *Ecological Applications* 17: 1279–1295.

Kušová, D., Bartoš, M. and Těšitel, J. 2002. Role of traditions in tourism development in the Czech part of the Bohemian Forest. *Silva Gabreta* 8: 265–274.

Kušová, D., Bartoš, M. and Těšitel, J. 2005. The media image of the relationship between nature protection and socio-economic development in selected protected landscape areas. *Silva Gabreta* 11: 123–133.

Kušová, D., Těšitel, J., Matějka, K. and Bartoš, M. 2008. Biosphere reserves: An attempt to form sustainable landscapes. A case study of three biosphere reserves in the Czech Republic. *Landscape and Urban Planning* 84: 38–51.

Kusumanto, Y. and Sirait, M.T. 2001. *Community Participation in Forest Resource Management in Indonesia: Policies, Practices, Constraints and Opportunities. Southeast Asia Policy Research Working Paper, No. 28*. International Council for Research in Agroforestry (ICRAF) Southeast Asia.

Kwan, M.-P. 2004. Beyond difference: From canonical geography to hybrid geographies. *Annals of the Association of American Geographers* 94: 756–763.

Ladle, R.J. and Whittaker, R.J. 2011. *Conservation Biogeography*. New Jersey: John Wiley & Sons.

Lane, M.B. 2001. Affirming new directions in planning theory: Comanagement of protected areas. *Society and Natural Resources* 14: 657–671.

Lawrence, A. 2008. Experiences with participatory conservation in post-socialist Europe. *The International Journal of Biodiversity Science and Management* 4: 179–186.

Lefebvre, H. 2001. *The Production of Space*. Oxford: Blackwell.

LeGates, R.T. and Stout, F. 2011. *City Reader*. London: Taylor & Francis.

Legislative Services Branch. 2013. Canada National Parks Act, http://laws-lois. justice.gc.ca/eng/acts/N-14.01/ (accessed January 6, 2014).

Lehmann, S. 1995. *Privatizing Public Lands*. Oxford: Oxford University Press.

Leib, J.I. 1998. Communities of interest and minority districting after Miller v. Johnson. *Political Geography* 17: 683–699.

Lekan, T. 2009. A 'Noble Prospect': Tourism, heimat, and conservation on the Rhine, 1880–1914. *The Journal of Modern History* 81: 824–858.

Lemieux, C.J. and Scott, D.J. 2005. Climate change, biodiversity conservation and protected area planning in Canada. *Canadian Geographer / Le Géographe canadien* 49: 384–397.

Lemieux, C.J., Beechey, T.J. and Gray, P.A. 2011. Prospects for Canada's protected areas in an era of rapid climate change. *Land Use Policy* 28: 928–941.

Librová, H. 1987. *Sociální Potřeba a Hodnota Krajiny*. Brno: UJEP.

Lindberg, K. and Johnson, R.L. 1997. Modeling resident attitudes toward tourism. *Annals of Tourism Research* 24: 402–424.

Liu, J. and Diamond, J. 2005. China's environment in a globalizing world. *Nature* 435: 1179–1186.

Liu, J., Ouyang, Z. and Miao, H. 2010. Environmental attitudes of stakeholders and their perceptions regarding protected area-community conflicts: A case study in China. *Journal of Environmental Management* 91: 2254–2262.

Löfgren, O. 1987. Rational and sensitive: Changing attitudes to time, nature and the home. In *Culture Builders: A Historical Anthropology of Middle-Class Life*, edited by J. Frykman and O. Löfgren. New Brunswick: Rutgers University Press.

Lorah, P. and Southwick, R. 2003. Environmental protection, population change, and economic development in the rural Western United States. *Population and Environment* 24: 255–272.

Lovejoy, T.E. 2002. Biodiversity: Threats and challenges. In *Biodiversity, Sustainability and Human Communities: Protecting Beyond The Protected*, edited by T. O'Riordan and S. Stoll-Kleemann. Cambridge: Cambridge University Press.

Lowry, W.R. 1994. Paved with political intentions: The impact of structure on the National Park services of Canada and the United States. *Policy Studies Journal* 22: 44–58.

Lupp, G., Höchtl, F. and Wende, W. 2011. 'Wilderness' – A designation for Central European landscapes? *Land Use Policy* 28: 594–603.

Manzo, L.C. and Perkins, D.D. 2006. Finding common ground: The importance of place attachment to community participation and planning. *Journal of Planning Literature* 20: 335–350.

Mappatoba, M. and Birner, R. 2004. *Co-management of Protected Areas: The Case of Community Agreements on Conservation in the Lore Lindu National Park, Central Sulawesi, Indonesia.* Germany: Cuvillier Verlag.

Martin, G.V., Kormos, F.C., Zunino, F., et al. 2008. Wilderness Momentum in Europe. *International Journal of Wilderness* 14: 34–43.

Martin, P.L. 2011. Global governance from the Amazon: Leaving oil underground in Yasuní National Park, Ecuador. *Global Environmental Politics* 11: 22–42.

Massey, D. 2001. Living in Wythenshawe. In *The Unknown City: Contesting Architecture and Social Space*, edited by I. Borden, J. Kerr, J. Rendell and A. Pivaro. Cambridge: MIT Press. 459–475.

Matagne, P. 1998. The politics of conservation in France in the 19th century. *Environment and History* 4: 359–367.

Mathur, P.K. and Sinha, P.R. 2008. Looking beyond protected area networks: A paradigm shift in approach for biodiversity conservation. *International Forestry Review* 10: 305–314.

Matsuoka, R.H. and Kaplan, R. 2008. People needs in the urban landscape: Analysis of landscape and urban planning contributions. *Landscape and Urban Planning* 84: 7–19.

McCarthy, J., Lloyd, G. and Illsley, B. 2002. National parks in Scotland: Balancing Environment and Economy. *European Planning Studies* 10: 665–670.

McCleave, J., Espiner, S. and Booth, K. 2006. The New Zealand people–park relationship: An exploratory model. *Society & Natural Resources* 19: 547–561.

McLean, J. and Straede, S. 2003. Conservation, relocation, and the paradigms of park and people management: A case study of Padampur villages and the Royal Chitwan National Park, Nepal. *Society and Natural Resources* 16: 509–526.

McNeely, J. A. 1994. Protected areas for the 21st century: Working to provide benefits to society. *Biodiversity & Conservation* 3: 390–405.

McNeely, J.A. and Miller, K. 1984. *National Parks, Conservation, and Development: The Role of Protected Areas in Sustaining Society: Proceedings of the World Congress on National Parks, Bali, Indonesia, 11–22 October 1982.* Washington, DC: Smithsonian Institution Press.

McShane, T.O. and Wells, M.P. 2004. Integrating protected area management with local needs and aspirations. *AMBIO: A Journal of the Human Environment* 33: 513–519.

Mehta, J.N. and Kellert, S.R. 1998. Local attitudes toward community-based conservation policy and programmes in Nepal: A case study in the Makalu-Barun Conservation Area. *Environmental Conservation* 25: 320–333.

Mels, T. 2002. Nature, home, and scenery: The official spatialities of Swedish national parks. *Environment and Planning D: Society and Space* 20: 135–154.

Merenlender, A.M., Huntsinger, L., Guthey, G. and Fairfax, S.K. 2004. Land trusts and conservation easements: Who is conserving what for whom? *Conservation Biology* 18: 65–76.

Meyer, T., Kiener, H. and Krenova, Z. 2009. The wild heart of Europe. *International Journal of Wilderness* 15: 33–40.

Miranda, L.M. and LaPalme, S. 1997. User rights and biodiversity conservation. In *Last Stand: Protected Areas and The Defense of Tropical Biodiversity*, edited by R.A. Kramer, C. van Schaik and J. Johnson. New York and Oxford: Oxford University Press. 133–156.

Mitchell, B. 2005. Editorial. *Parks: The International Journal for Protected Area Managers* 15: 1–5.

Mize, J. 2006. Integrating indigenous cultural traditions in the management of protected marine resources – the Fiordland Example, http://works.bepress.com/james_mize/5 (accessed January 6, 2014).

Molloy, L., Reedy, M., Watson, A.E., et al. 2000. Wilderness within World heritage: Te Wahipounamu, New Zealand. In *AE Watson, GH Aplet and JC Hendee (Compilers) Personal, Societal, and Ecological Values of Wilderness: Sixth World Wilderness Congress Proceedings on Research, Management, and Allocation.* 162–167, http://www.fs.fed.us/rm/pubs/rmrs_p014/rmrs_p0 14_162_167.pdf (accessed January 6, 2014).

Morrill, R.L. 1987. Redistricting, region and representation. *Political Geography Quarterly* 6: 241–260.

Mulder, M.B. and Coppolillo, P. 2005. *Conservation: Linking Ecology, Economics, and Culture*. New Jersey: Princeton University Press.

Munck, R. 2005. Neoliberalism and politics, and the politics of neoliberalism. In *Neoliberalism: A Critical Reader*, edited by A. Saad-Filho and D. Johnston. London: Pluto. 60–69.

Mundilová, E. 2007. *Turismus, Lázeňství a Environmentální Management v CHKO Slavkovský Les – Postoje a Názory Místních Obyvatel*. Prague: ÚŽP PřF UK.

Murray, G. and King, L. 2012. First Nations Values in Protected Area Governance: Tla-o-qui-aht Tribal Parks and Pacific Rim National Park Reserve. *Human Ecology* 40: 385–395.

Národní Park Šumava. 2010. Správa Národního parku a chráněné krajinné oblasti Šumava, http://www.npsumava.cz.

Nastov, A. 2000. *Report of Nature Conservation in The Former Republic of Macedonia, Convention on the Conservation of European Wildlife and Natural Habitats*. Strasbourg: Council of Europe.

Nastov, A. and Micevski, B. 1994. *National Report of Status of Conservation of Biological Diversity in Protected Areas of the Republic of Macedonia*. Geneva and Skopje: UNEP.

Nelson, F. 2012. *Community Rights, Conservation and Contested Land: The Politics of Natural Resource Governance in Africa*. London: Routledge.

Nepal, S.K. 2002. Involving indigenous peoples in protected area management: comparative perspectives from Nepal, Thailand, and China. *Environmental management* 30: 748–763.

Neumann, R. 1992. The political ecology of wildlife conservation in the Mount Meru Area, Northeast Tanzania. *Land Degradation and Rehabilitation* 3: 85–98.

Neumann, R.P. 1998. *Imposing Wilderness: Struggles over Livelihood and Nature Preservation in Africa*. Berkeley: University of California Press.

Neves, K. and Igoe, J. 2012. Uneven development and accumulation by dispossession in nature conservation: Comparing recent trends in the Azores and Tanzania. *Tijdschrift voor economische en sociale geografie* 103: 164–179.

Nováková, E. 2004. *Vnímání Dopadů Cestovního Ruchu Rezidenty Českého Ráje*. Prague: KSGRR, PřF UK, Diplomová práce.

Obua, J. and Harding, D.M. 1996. Visitor characteristics and attitudes towards Kibale National Park, Uganda. *Tourism Management* 17: 495–505.

Oguz, D. 2000. User surveys of Ankara's urban parks. *Landscape and Urban Planning* 52: 165–171.

Oku, H. and Fukamachi, K. 2006. The differences in scenic perception of forest visitors through their attributes and recreational activity. *Landscape and Urban Planning* 75: 34–42.

Ostrom, E., Gardner, R. and Walker, J. 1994. *Rules, Games and Common-Pool Resources*. Ann Arbor: University of Michigan Press.

Oszlányi, J., Grodzińska, K., Badea, O. and Shparyk, Y. 2004. Nature conservation in Central and Eastern Europe with a special emphasis on the Carpathian Mountains. *Environmental Pollution* 130: 127–134.

Papageorgiou, K. and Vogiatzakis, I.N. 2006. Nature protection in Greece: An appraisal of the factors shaping integrative conservation and policy effectiveness. *Environmental Science and Policy* 9: 476–486.

Parks Canada Agency. 2000. *Unimpaired for Future Generations? Volume 1: A Call for Action*. Ottawa: Parks Canada Agency.

Parks, S.A. and Harcourt, A.H. 2002. Reserve size, local human density, and mammalian extinctions in U.S. protected areas. *Conservation Biology* 16: 800–808.

Parry, M.L. 2007. *Climate Change 2007: Impacts, Adaptation and Vulnerability: Contribution of Working Group II to the Fourth Assessment Report of the Intergovernmental Panel on Climate Change*. Cambridge: Cambridge University Press.

Pauchard, A. and Villarroel, P. 2002. Protected areas in Chile: History, current status, and challenges. *Natural Areas Journal* 22: 318–330.

Pavlikakis, G.E. and Tsihrintzis, V.A. 2006. Perceptions and preferences of the local population in Eastern Macedonia and Thrace National Park in Greece. *Landscape and Urban Planning* 1–16.

Pavlínek, P. and Pickles, J. 2004. Environmental pasts/environmental futures in post-socialist Europe. *Environmental Politics* 13: 237–265.

Payton, M.A., Fulton, D.C. and Anderson, D.H. 2005. Influence of place attachment and trust on civic action: A study at Sherburne National Wildlife Refuge. *Society & Natural Resources* 18: 511–528.

Peck, J., Theodore, N. and Brenner, N. 2010. Postneoliberalism and its malcontents. *Antipode* 41: 1236–1258.

Peres, C.A. 2000. Evaluating the impact and sustainability of subsistence hunting at multiple Amazonian forest sites. In *Hunting for Sustainability in Tropical Forests*, edited by J.G. Robinson and E.L. Bennett. New York: Columbia University Press.

Peters, R.L. 1985. The greenhouse effect and nature reserves. *BioScience* 35: 707–717.

Petrosillo, I., Zurlini, G., Corliano, M.E., et al. 2007. Tourist perception of recreational environment and management in a marine protected area. *Landscape and Urban Planning* 79: 29–37.

Petrova, S., Bouzarovski, S. and Čihař, M. 2009a. From inflexible national legislation to flexible local governance: Management practices in the Pelister National Park, Republic of Macedonia. *GeoJournal* 74: 589–598.

Petrova, S., Čihař, M. and Bouzarovski, S. 2009b. Conservationist or fashionista? Urban dwellers' expectations from national parks in the Republic of Macedonia. *Urbani Izziv* 128–135.

Petrova, S., Čihař, M. and Bouzarovski, S. 2011. Local nuances in the perception of nature protection and place attachment: A tale of two parks. *Area* 43: 327–335.

Petrova, S., Posová, D., House, A. and Sýkora, L. 2013. Discursive framings of low carbon urban transitions: The contested geographies of 'satellite settlements' in the Czech Republic. *Urban Studies* 50: 1439–1455.

Phillips, A. 2004. The history of the international system of protected area management categories. *Parks: The international journal for protected area managers* 14: 4–15.

Piermattei, S. 2013. Local farmers vs. environmental universalism: Conflicts over nature conservation in the Parco Nazionale dei Monti Sibillini, Italy. *Ecology* 20: 255–341.

Pimbert, M.P. and Pretty, J.N. 1995. *Parks, People and Professionals: Putting 'Participation' into Protected Area Management. Discussion Paper No 57.* Geneva: United Nations Research Institute for Sustainable Development.

Plesník, J. and Roudná, M. 2000. *Status of Biological Resources and Implementation of the Convention on Biological Diversity in the Czech Republic.* Prague: Ministry of the Environment of the Czech Republic, First Report.

Poputoaia, D. and Bouzarovski, S. 2010. Regulating district heating in Romania: Legislative challenges and energy efficiency barriers. *Energy Policy* 38: 3820–3829.

Pretty, J. 1995. The many interpretations of participation. *In Focus* 16: 4–5.

Proshansky, H.M., Fabian, A.K. and Kaminoff, R. 1983. Place-identity: Physical world socialization of the self. *Journal of Environmental Psychology* 3: 57–83.

Pucher, J. and Buehler, R. 2005. Transport policies in central and Eastern Europe. *Transport strategy, policy, and institutions*, http://policy.rutgers.edu/faculty/ pucher/PDF%20of%20chapter.pdf (accessed January 6, 2014).

Pullin, A.S., Báldi, A., Can, O.E., et al. 2009. Conservation focus on Europe: Major conservation policy issues that need to be informed by conservation science. *Conservation Biology* 23: 818–824.

Putnam, R.D. 1995. Bowling alone: America's declining social capital. *Journal of Democracy* 6: 65–78.

Putnam, R.D. 2000. *Bowling Alone: The Collapse and Revival of American Community*. New York: Simon & Schuster.

Radeloff, V.C., Stewart, S.I., Hawbaker, T.J., et al. 2010. Housing growth in and near United States protected areas limits their conservation value. *Proceedings of the National Academy of Sciences* 107: 940–945.

Ranciere, J. 2010. *Dissensus: On Politics and Aesthetics*. Continuum.

Rao, K.S., Maikhuri, R.K., Nautiyal, S. and Saxena, K.G. 2002. Crop damage and livestock depredation by wildlife: A case study from Nanda Devi Biosphere Reserve. *Journal of Environmental Management* 66: 317–327.

Reid, H. 2001. Contractual national parks and the Makuleke community. *Human Ecology* 29: 135–155.

Reynolds, G. and Elson, M.J. 1996. The sustainable use of sensitive countryside sites for sport and active recreation. *Journal of Environmental Planning and Management* 39: 563–576.

Robbins, P. 2004. *Political Ecology: A Critical Introduction 'Why Bother to Argue that Nature (or Forest or Land Degradation ...) is Constructed?'* Oxford: Blackwell.

Robins, S. and Waal, K. van der. 2008. 'Model tribes' and iconic conservationists? The Makuleke restitution case in Kruger National Park. *Development and Change* 39: 53–72.

Robinson, J.G. and Bennett, E.L. 2000. Carrying capacity limits to sustainable hunting in tropical forests. In *Hunting for Sustainability in Tropical Forests*, edited by J.G. Robinson and E.L. Bennett. New York: Columbia University Press.

Rodela, R. and Udovč, A. 2008. Participation in nature protection: Does it benefit the local community? A Triglav National Park case study. *International Journal of Biodiversity Science, Ecosystem Services & Management* 4: 209–218.

Rose, N. 1999. *Powers of Freedom: Reframing Political Thought*. Cambridge: Cambridge University Press.

Rosenzweig, L.M. 2003. *Win-Win Ecology: How the Earth's Species Can Survive in the Midst of Human Enterprise*. Oxford: Oxford University Press.

Rugg, D.S. 1994. Communist legacies in the Albanian landscape. *Geographical Review* 84: 59.

Runte, A. 1997. *National Parks: The American Experience*. Nebraska: University of Nebraska Press, http://www.cr.nps.gov/history/online_books/runte1/index. htm.

Rytteri, T. and Puhakka, R. 2009. Formation of Finland's national parks as a political issue. *Ethics, Place & Environment* 12: 91–106.

Salafsky, N. and Wollenberg, E. 2000. Linking livelihoods and conservation: A conceptual framework and scale for assessing the integration of human needs and biodiversity. *World Development* 28: 1421–1438.

Sandom, C., Bull, J., Canney, S. and Macdonald, D.W. 2012. Exploring the value of wolves (*Canis lupus*) in landscape-scale fenced reserves for ecological restoration in the Scottish Highlands. In *Fencing for Conservation*, edited by M.J. Somers and M. Hayward. New York: Springer. 245–276, http:// link.springer.com/chapter/10.1007/978-1-4614-0902-1_14 (accessed January 5, 2014).

Sarre, P. and Jehlička, P. 2007. Environmental movements in space-time: The Czech and Slovak republics from Stalinism to post-socialism. *Transactions of the Institute of British Geographers* 32: 346–362.

Schelhaas, M.-J., Nabuurs, G.-J. and Schuck, A. 2003. Natural disturbances in the European forests in the 19th and 20th centuries. *Global Change Biology* 9: 1620–1633.

Schwartz, K.Z.S. 2006. *Nature and National Identity after Communism: Globalizing the Ethnoscape*. Pittsburgh, PA: University of Pittsburgh Press.

Scott, D., Malcolm, J.R. and Lemieux, C. 2002a. Climate change and modelled biome representation in Canada's national park system: Implications for system planning and park mandates. *Global Ecology and Biogeography* 11: 475–484.

Scott, D., Malcolm, J.R. and Lemieux, C. 2002b. Climate change and modelled biome representation in Canada's national park system: Implications for system planning and park mandates. *Global Ecology and Biogeography* 11: 475–484.

Secretariat of the Convention on Biological Diversity. 2001. *Handbook of the Convention on Biological Diversity*. London: Earthscan.

Sellars, R.W. 1997. *Preserving Nature in the National Parks: A History*. Yale University Press.

Sherman, S.A., Varni, J.W., Ulrich, R.S. and Malcarne, V.L. 2005. Post-occupancy evaluation of healing gardens in a pediatric cancer center. *Landscape and Urban Planning* 73: 167–183.

Shultis, J. and More, T. 2011. American and Canadian national park agency responses to declining visitation. *Journal of Leisure Research* 43: 110–132.

Shumaker, S.A. and Taylor, R. 1983. Toward a classification of people-place relationships: A model of attachment to place. In *Environmental Psychology: Directions and Perspectives*, edited by N.R. Feimer and E.S. Geller. New York: Praeger. 219–251.

Smaldone, D., Harris, C.C., Sanyal, N. and Lind, D. 2005. Place attachment and management of critical park issues in Grand Teton National Park. *Journal of Park and Recreation Administration* 23: 90–114.

Smurr, R.W. 2008. Lahemaa: The paradox of the USSR's first national park. *Nationalities Papers* 36: 399–423.

Sorrenson, M.P.K. 1968. *Origins of European settlement in Kenya*. Oxford: Oxford University Press.

Spinage, C. 1998. Social change and conservation misrepresentation in Africa. *Oryx* 32: 265–276.

Stabler, M.J., Papatheodorou, A. and Sinclair, M.T. 2009. *The Economics of Tourism*. London: Routledge.

Staddon, C. 2009. Towards a critical political ecology of human–forest interactions: Collecting herbs and mushrooms in a Bulgarian locality. *Transactions of the Institute of British Geographers* 34.

Stahl, J. 2010. The rents of illegal logging: The mechanisms behind the rush on forest resources in Southeast Albania. *Conservation and Society* 8: 140–150.

Staple, T. and Wall, G. 1996. Climate change and recreation in Nahanni National Park Reserve. *Canadian Geographer* [*Le Géographe canadien*] 40: 109–120.

Stark, D. 1996. Recombinant property in East European capitalism. *American Journal of Sociology* 101: 993–1027.

Stegner, W.E. 1987. *The American West as Living Space*. Ann Arbor, MI: University of Michigan Press.

Stein, B.A., Staudt, A., Cross, M.S., et al. 2013. Preparing for and managing change: Climate adaptation for biodiversity and ecosystems. *Frontiers in Ecology and the Environment* 11: 502–510.

Sterl, P., Brandenburg, C. and Arnberger, A. 2008. Visitors' awareness and assessment of recreational disturbance of wildlife in the Donau-Auen National Park. *Journal for Nature Conservation* 16: 135–145.

Stevens, S. 1997. The legacy of Yellowstone. In *Conservation through Cultural Survival: Indigenous Peoples and Protected Areas*, edited by S. Stevens. Washington, DC: Island Press. 13–32.

Stokols, D. and Shumaker, S. A. 1981. People in places: A transactional view of settings. In *Cognition, Social Behavior, and the Environment*, edited by D. Harvey. Hillsdale: Erlbaum. 441–488.

Stringer, L.C., Dougill, A.J., Fraser, E., et al. 2006. Unpacking 'participation' in the adaptive management of social–ecological systems: A critical review. *Ecology and Society* 11: 39.

Suckall, N., Fraser, E.D.G., Cooper, T. and Quinn, C. 2009. Visitor perceptions of rural landscapes: A case study in the Peak District National Park, England. *Journal of Environmental Management* 90: 1195–1203.

Švajda, J. 2008. Participatory conservation in a post-communist context: The Tatra National Park and Biosphere Reserve, Slovakia. *International Journal of Biodiversity Science, Ecosystem Services & Management* 4: 200–208.

Swedish Institute. 1990. *The Saami People in Sweden*. Stockholm: The Swedish Institute.

Swyngedouw, E.A. 2007. Impossible 'sustainability' and the postpolitical condition. In *The Sustainable Development Paradox: Urban Political Economy in the United States and Europe*, edited by R. Krueger and D. Gibbs. New York: Guilford Press. 13–40.

Taff, G.N., Müller, D., Kuemmerle, T., et al. 2010. Reforestation in Central and Eastern Europe after the Breakdown of Socialism. In *Reforesting Landscapes*, edited by H. Nagendra and J. Southworth. Netherlands: Springer. 121–147, http://link.springer.com/chapter/10.1007/978-1-4020-9656-3_6 (accessed January 6, 2014).

Terborgh, J. 2004. *Requiem for Nature*. Washington, DC: Island Press.

Terborgh, J. and van Schaik, C. 2002. Why the World needs parks. In *Making Parks Work: Strategies for Preserving Tropical Nature*, edited by J. Terborgh, C. van Schaik, L. Davenport and M. Rao. Washington, DC: Island Press.

Terborgh, J., Estes, J.A., Paquet, P., et al. 1999. The role of top carnivores in regulating terrestrial ecosystems. In *Continental Conservation: Scientific Foundations of Regional Reserve Networks*, edited by M.E. Soule and J. Terborgh. Washington, DC: Island Press.

Theil, S. 2005. Into the woods: Economics and declining birthrates are pushing large swaths of Europe back to their primeval state, with wolves taking the place of people. *Newsweek*.

Tickle, A. 2000. Regulating environmental space in socialist and post-socialist systems: Nature and landscape conservation in the Czech Republic. *Journal of Contemporary European Studies* 8: 57–78.

Tomićević, J., Shannon, M.A. and Milovanović, M. 2010. Socio-economic impacts on the attitudes towards conservation of natural resources: Case study from Serbia. *Forest Policy and Economics* 12: 157–162.

Tönnies, F. 2011. *Community and Society*. New York: Courier Dover Publications.

Trakolis, D. 2001. Local residents's perceptions of planning and management issues in Prespes Lakes National Park, Greece. *Journal of Environmental Management* 61: 227–241.

Treasury, G.B. 2007. *The Economics of Climate Change: The Stern Review*. Cambridge: Cambridge University Press.

Ulrich, R.S., Simons, R.F., Losito, B.D., et al. 1991. Stress recovery during exposure to natural and urban environments. *Journal of environmental Psychology* 11: 201–230.

UNESCO. 2006. *Sweden, Laponian Area. Summary of the Periodic Report on the State of Conservation*. Paris: UNESCO.

Ürge-Vorsatz, D., Miladinova, G. and Paizs, R. 2006. Energy in transition: From the iron curtain to the European Union. *Energy Policy* 27: 2279–2297.

US Congress. 1964. *The Wilderness Act*. Washington, DC.

Vacek, S. and Podrázský, V.V. 2003. Forest ecosystems of the Šumava Mts. and their management. *Journal of Forest Science* 49: 291–301.

Van Amerom, M. and Büscher, B. 2005. Peace parks in Southern Africa: Bringers of an African Renaissance? *The Journal of Modern African Studies* 43: 159–182.

Van Schaik, C. and Rijksen, H.D. 2002. Integrated conservation and development projects: Problems and potentials. In *Making Parks Work: Strategies for Preserving Tropical Nature*, edited by J. Terborgh, C. van Schaik, L. Davenport and M. Rao. Washington, DC: Island Press.

Villarroel, P. 1996. Efecto del turismo en el desarrollo local. *Ambiente y Desarrollo* 58–64.

Von Ruschkowski, E. and Mayer, M. 2011. From conflict to partnership? Interactions between protected areas, local communities and operators of tourism enterprises in two German national park regions. *Journal of Tourism and Leisure Studies* 17: 147–181.

Vorkinn, M. and Riese, H. 2001. Environmental concern in a local context: The significance of place attachment. *Environment and Behavior* 33: 249–263.

Wade, A.A. and Theobald, D.M. 2010. Residential development encroachment on U.S. protected areas. *Conservation Biology* 24: 151–161.

Waithaka, J. 2012. The Kenya wildlife service in the 21st century: Protecting globally significant areas and resource. *The George Wright Society Journal of Parks, Protected Areas and Cultural Sites* 29: 21–29.

Walker, B., Carpenter, S., Anderies, J., et al. 2002. Resilience management in social-ecological Systems: A working hypothesis for a participatory approach. *Conservation Ecology* 6.

Wallner, A., Bauer, N. and Hunziker, M. 2007. Perceptions and evaluations of biosphere reserves by local residents in Switzerland and Ukraine. *Landscape and Urban Planning* 83: 104–114.

Walters, C.J. and Holling, C.S. 1990. Large-scale management experiments and learning by doing. *Ecology* 71: 2060.

Wang, G., Innes, J.L., Wu, S.W., et al. 2012. National park development in China: Conservation or commercialization? *AMBIO* 41: 247–261.

Watson, E.M.J., Cross, M., Rowland, E., et al. 2011. Planning for species conservation in a time of climate change. In *Climate Change: Research and Technology for Adaptation and Mitigation*, edited by J. Blanco. Rijeka: InTech. 379–402.

Watters, E. 2003. *Urban Tribes: A Generation Redefines Friendship, Family, and Commitment*. London and New York: Bloomsbury.

Watts, M. 1998. Recombinant capitalism: State, de-collectivisation and the agrarian question in Vietnam. In *Theorising Transition. The Political Economy of Post-Communist Transformation*, edited by A. Pickles and A. Smith. London: Routledge. 450–505.

Watts, M.J. 2004. Antinomies of community: Some thoughts on geography, resources and empire. *Transactions of the Institute of British Geographers* 29: 195–216.

Wells, M.P. and Brandon, K. 1992. *People and Parks: Linking Protected Area Management with Local Communities*. Washington, DC: The World Bank. Agency for International Development and World Wildlife Fund.

Wells, M.P. and Brandon, K. 1993. The principles and practice of buffer zones and local participation in biodiversity conservation. *Ambio* 22: 157–162.

Wesołowski, T. 2005. Virtual conservation: How the European Union is turning a blind eye to its vanishing primeval forests. *Conservation Biology* 19: 1349–1358.

Wesołowski, T. 2007. Primeval conditions – what can we learn from them? *Ibis* 149: 64–77.

Western, D. and Waithaka, J. 2005. Policies for reducin human-wildlife conflict: A Kenya case study. In *People and Wildlife, Conflict Or Co-existence?* edited by R. Woodroffe, Simon Thirgood and A. Rabinowitz. Cambridge: Cambridge University Press. 357–372.

Westman, W.E. 1990. Park management of exotic plant species: Problems and issues. *Conservation Biology* 4: 251–260.

Williams, B.K. 2011. Adaptive management of natural resources – framework and issues. *Journal of Environmental Management* 92: 1346–1353.

Williams, D.R. and Roggenbuck, J.W. 1990. A framework for examining the meaning of recreation places: Place attachment, mode of experience, and environmental dispositions. In *Proceedings: The Third Symposium on Social Science in Resource Management,* edited by J.H. Gramann. College Station: Texas A&M University. 70–72.

Williams, R. 1972. Ideas of nature. In *Ecology: The Shaping Enquiry*, edited by J. Benthall. London: Longman. 146–164.

Wilshusen, P.R., Brechin, S.R., Fortwangler, C.L. and West, P.C. 2002. Reinventing a square wheel: Critique of a resurgent 'Protection paradigm' in international biodiversity conservation. *Society & Natural Resources* 15: 17–40.

Winchester, H.P.M. 1999. Interviews and questionnaires as mixed methods in population geography: The case of lone fathers in Newcastle, Australia. *The Professional Geographer* 51: 60–67.

Wolmer, W. 2003. Transboundary Protected Area governance: tensions and paradoxes. In *Prepared for the workshop on Transboundary Protected Areas in the Governance Stream of the 5th World Parks Congress, Durban, South Africa.* 12–13, http://www.tbpa.net/docs/WPCGovernance/WilliamWolmer.pdf (accessed January 6, 2014).

World Rainforest Movement. 2011. Letter from the State of Acre – In defence of life and the integrity of the peoples and their territories against REDD and the commodification of nature. *WRM English*, http://wrm.org.uy/other-relevant-information/letter-from-the-state-of-acre-in-defence-of-life-and-the-integrity-of-the-peoples-and-their-territories-against-redd-and-the-commodification-of-nature/ (accessed January 6, 2014).

Xu, J. and Melick, D.R. 2007. Rethinking the effectiveness of public protected areas in Southwestern China. *Conservation Biology* 21: 318–328.

Young, I.M. 1990. The ideal of community and the politics of difference. In *Feminism/Postmodernism*, edited by J.L. Nicholson. New York: Routledge. 300–323.

Young, J., Richards, C., Fischer, A., et al. 2007. Conflicts between biodiversity conservation and human activities in the Central and Eastern European countries. *Ambio* 36.

Zeide, B. 2001. Thinning and growth: A full turnaround. *Journal of Forestry* 99: 20–25.

Zimmerer, K.S., Galt, R.E. and Buck, M.V. 2004. Globalization and multi-spatial trends in the coverage of protected-area conservation (1980–2000). *Ambio* 33: 520–529.

Zunino, F. 1980. *'Wilderness': Una nuova esigenza di conservazione delle aree naturali*. Roma: Ministero dell'Agricoltura e delle Foreste.

Zvára, K. 2003. *Biostatistika*. Praha: Nakladatelství Karolinum.

Index